Green Energy and Technology

More information about this series at http://www.springer.com/series/8059

José Luz Silveira
Editor

Sustainable Hydrogen Production Processes

Energy, Economic and Ecological Issues

Springer

Editor
José Luz Silveira
Energy Department GOSE
College of Engineering of Guaratinguetá - UNESP
Guaratinguetá, São Paulo
Brazil

ISSN 1865-3529 ISSN 1865-3537 (electronic)
Green Energy and Technology
ISBN 978-3-319-41614-4 ISBN 978-3-319-41616-8 (eBook)
DOI 10.1007/978-3-319-41616-8

Library of Congress Control Number: 2016952244

Printed on acid-free paper

This Springer imprint is published by Springer Nature
The registered company is Springer International Publishing AG Switzerland

Acknowledgements

We would like to express our warmest thanks to all those who have contributed to publishing this book, especially all of its authors and coauthors, including the FAV Language Services representative who had made a valiant effort to translate, improve, and revise the language style of chapters which have not been written by native speakers.

Contents

Contributors

Fernando Henrique Mayworm de Araujo Group of Optimization of Energy Systems—GOSE, College of Engineering of Guaratinguetá, Institute of Bioenergy Research—IPBEN, São Paulo State University—UNESP, Guaratinguetá, SP, Brazil

Lúcia Bollini Braga Group of Optimization of Energy Systems—GOSE, College of Engineering of Guaratinguetá, Institute of Bioenergy Research—IPBEN, São Paulo State University—UNESP, Guaratinguetá, SP, Brazil

Túlio Stefani Colombaroli Group of Optimization of Energy Systems—GOSE, College of Engineering of Guaratinguetá, Institute of Bioenergy Research—IPBEN, São Paulo State University—UNESP, Guaratinguetá, SP, Brazil

Daniel Travieso Pedroso Group of Optimization of Energy Systems—GOSE, College of Engineering of Guaratinguetá, Institute of Bioenergy Research—IPBEN, São Paulo State University—UNESP, Guaratinguetá, SP, Brazil

Márcio Evaristo da Silva Group of Optimization of Energy Systems—GOSE, College of Engineering of Guaratinguetá, Institute of Bioenergy Research—IPBEN, São Paulo State University—UNESP, Guaratinguetá, SP, Brazil

José Luz Silveira Group of Optimization of Energy Systems—GOSE, College of Engineering of Guaratinguetá, Institute of Bioenergy Research—IPBEN, São Paulo State University—UNESP, Guaratinguetá, SP, Brazil

Luis Carlos Félix Tapia Department of Industrial Ecology, Royal Institute of Technology (KTH), Stockholm, Sweden

Celso Eduardo Tuna Group of Optimization of Energy Systems—GOSE, College of Engineering of Guaratinguetá, Institute of Bioenergy Research—IPBEN, São Paulo State University—UNESP, Guaratinguetá, SP, Brazil

Lucas Fachini Vane Group of Optimization of Energy Systems—GOSE, College of Engineering of Guaratinguetá, Institute of Bioenergy Research—IPBEN, São Paulo State University—UNESP, Guaratinguetá, SP, Brazil

Rolando Zanzi Vigouroux Department of Industrial Ecology, Royal Institute of Technology (KTH), Stockholm, Sweden

Abbreviations

A	Cross sectional area (m^2)
ADP	Adenosine diphosphate (–)
ANEEL	Brazilian Electricity Regulatory Agency (–)
$Al(OH)_3$	Aluminum hydroxide (–)
ATP	Adenosine triphosphate (–)
Al_2O_3	Alumina (–)
B	Boron (–)
C	Carbon (–)
Ca	Calcium (–)
$CaCl_2$	Calcium chloride (–)
CC	Direct current (–)
C_{H_2O}	Water cost (US$/$m^3$)
Cmain	Maintenance cost (US$/kWh)
Cgrowthmedium	Cost of algae growth medium added at every 164 h (US$/year)
Cproductmedium	Cost of replenishing the algae production medium at every 164 h (US$/year)
C_nH_m	Hydrocarbon (–)
C_2H_6	Ethane (–)
CH_4	Methane (–)
$C_6H_{12}O_6$	Glucose (–)
C_2H_5OH	Ethanol (–)
CO	Carbon monoxide (–)
Co	Cobalt (–)
CO_2	Carbon dioxide (–)
$(CO_2)_e$	Equivalent carbon dioxide (kg/kg_{comb})
Cop	Operating cost of the reformer (US$/kWh)
Cu	Copper (–)
e^-	Electron (–)
e_{adapt}	Energy required for the period of algae adaptation to an anaerobic medium (kJ)

E_{FUEL}	Energy supplied by the fuel (kW)
Econs	Energy consumption (kW)
P_{steam}	Steam power (kW)
$e_{grow.}$	Energy required for the period of algae growth (kJ)
$E_{Ene_{(sup)}}$	Required energy for the electrolyzer (kW)
$E_{H_{2year}}$	Energy stored from hydrogen production (kWh/year)
$E_{hydrogen}$	Energy stored from hydrogen production (kJ)
$e_{prod.\ hydro}$	Required energy in the hydrogen production (kJ)
f	Annuity factor (1/year)
Fd	Ferredoxin (–)
Fe	Iron (–)
Fe_2O_3	Ferric oxide (–)
GWP	Global warming potential ($kgCO_{2(eq)}$/MWh)
h	Equivalent utilization period (h/year)
H	Hydrogen (–)
H^+	Hydrogen cation (–)
H_2	Hydrogen (–)
H_2ase	Hydrogenase (–)
H_2O	Water (–)
Inv	Investment cost (US$)
k	Payback period (years)
K	Equilibrium constant (–)
Kc	Equilibrium constant in terms of concentration (–)
Kp	Equilibrium constant in terms of partial pressure (–)
K_2HPO_4	Potassium hydrogen phosphate (–)
KH_2PO_4	Potassium dihydrogen phosphate (–)
KOH	Potassium hydroxide (–)
La_2O_3	Lanthanum (III) oxide (–)
$\dot{m}$	Mass flow (kg/h)
Mg	Magnesium (–)
MgO	Magnesium oxide (–)
$MgSO_4$	Magnesium sulfate (–)
Mn	Manganese (–)
Mo	Molybdenum (–)
MP	Particulate matter (–)
n	Number of moles (mol)
N	Nitrogen (–)
NADP	Nicotinamide adenine dinucleotide phosphate (–)
NaOH	Sodium hydroxide (–)
NG	Natural gas (–)
NH_4Cl	Ammonium chloride (–)
Ni	Nickel (–)
NOx	Nitrogen oxide (–)
O^{2-}	Oxygen ion (–)

O_2	Oxygen (–)
OH^-	Hydroxyl ion (–)
P	Pressure (atm)
SHPP	Small hydroelectric power plant (–)
LHV	Lower heating value (kJ/kg)
C_{Fuel}	Fuel cost (US$/kWh)
Pd	Palladium (–)
WP_{avai}	Wind power available (W)
C_{Elect}	Cost of electricity (US$/kWh)
PSI	Photosystem 1 (–)
PSII	Photosystem 2 (–)
Pt	Platinum (–)
Qi	Calorific power of the fuel (MJ/kg_{fuel})
r	Annual interest rate (%)
R	Universal gas constant (J/mol K)
Rh	Rhodium (–)
S	Sulfur (–)
Si	Silicon (–)
SiO_2	Silicon dioxide (–)
SO_2	Sulfur dioxide (–)
T	Temperature (K)
TAP	Tris-acetate-phosphate (–)
TAPS	Tris-acetate-phosphate without sulfur (–)
TiO_2	Titanium dioxide (–)
v	Wind speed (m/s)
Zn	Zinc (–)
Zr	Zirconium (–)
ZrO_2	Zirconium(II)oxide (–)
α	Degree of advancement (–)
ΔG	Gibbs free energy change (kJ/mol)
ΔH	Enthalpy variation (kJ/mol$_)$
Δh_{vapor}	Steam enthalpy variation (kJ/kg)
ΔS	Entropy variation (kJ/mol K)
ε	Ecological efficiency (%)
η_X	Efficiency of system X (%)
Π_g	Pollutant indicator (kg/MJ)
ρ_{air}	Air density (kg/m^3)
DEFC	Direct ethanol fuel cell
DMFC	Direct methanol fuel cell
FC	Fuel cell (Hydrogen)
FFV	Flex fuel vehicle
GHG	Green house gas
GOSE	Group of Optimization of Energy Systems at UNESP, Guaratinguetá

HBP	Hydrogen from biological processes
HCGCC	Hydrogen from coal gasification with carbon capture
HEPRS	Hydrogen from electrolysis powered by renewable sources
HSRE	Hydrogen from the steam reforming of ethanol
HSRNG	Hydrogen from the steam reforming of natural gas
ICE	Internal combustion engine
KOH	Potassium hydroxide
LCA	Life cycle analysis
LV	Light vehicle (motor vehicles that do not exceed 3.5 tons in gross weight)
MCA	Multicriteria analysis
PM_{10}	Particulate matter with diameter size that is no greater than 10 μm
PEMFC	Proton exchange membrane fuel cell
PPB	Part per billion
PTE	Potential to emit
R&D	Research and development

Chapter 1
Introduction

José Luz Silveira

1.1 Initial Considerations

The search for energy alternatives to replace or supplement the sources that are used nowadays has been a constant challenge for scholars and researchers. In this context, hydrogen has been researched, due to its large energy storage capacity and to reducing greenhouse gas emissions. Although it cannot be found in nature in its raw state, it is present in other elements like water, fossil fuels, and in all living things.

The feasibility and selection of a hydrogen production process is predicated on several factors, among which are: the scale of production, raw materials availability, and immediate use or storage (Fatsikostas et al. 2002).

Among the commercial targets of hydrogen production for energy generation purposes, the fuel cell stands out as the most prominent method. Such a configuration allows a decentralized energy generation, besides only releasing steam afterwards, thus increasing the energy supply with low environmental impact (Silva 1991).

The study of some hydrogen production processes is going to be analyzed in this work, in order to determine which process is the most technologically, economically, and ecologically feasible.

J.L. Silveira (✉)
Group of Optimization of Energy Systems—GOSE, College of Engineering of Guaratinguetá, Institute of Bioenergy Research—IPBEN, São Paulo State University—UNESP, Dr. Ariberto Pereira da Cunha Ave, 333. Pedregulho, Guaratinguetá 12516-410, São Paulo, Brazil
e-mail: joseluz@feg.unesp.br

J.L. Silveira (ed.), *Sustainable Hydrogen Production Processes*, Green Energy and Technology, DOI 10.1007/978-3-319-41616-8_1

1.2 Book Structure

This book is divided into six chapters, as described below:

Chapter 1 presents the initial considerations of studies conducted and the delimitation of the subject, as well as the structure and purpose of the present work.

Chapter 2 discusses the state-of-the-art in terms of hydrogen production processes. At first, the steam reforming reactions of ethanol, biogas, and natural gas are introduced. A study on the catalysts being used in the selected reforming processes is also presented with their respective operating temperatures, feed molar ratio, and conversion rates of reagents. Afterwards, the electrolysis process and types of electrolyzers are presented with the renewable energy sources required for such, as well as the settings adopted for this type of hydrogen production processes. This chapter is concluded with the presentation of the biological hydrogen production process from green algae, with the description of the alga strain and methodology for determining the amount of hydrogen produced.

In Chap. 3, thermodynamic studies are conducted for determining the energy efficiencies of each type of hydrogen production process. In the case of the steam reforming processes, a physicochemical analysis was previously conducted, which was based on the concepts of Gibbs free energy, Equilibrium constant, and Degree of advancement. In light of pressure and temperature conditions, the energy efficiency levels of such processes are determined. In the case of hydrogen production from renewable electrolytic processes, it was based on the electrolyzer's efficiency and the average efficiencies of wind, photovoltaic, and hydroelectric power plants. In the case of algae, it was considered the energy contained in the hydrogen being produced and the energy consumption levels during the periods of growth, adaptation, and hydrogen production.

In Chap. 4, it is determined the cost of hydrogen production per kWh of electricity produced (US$/kWh) for each process, and a study drawing a comparison of their cost is thereof presented on the basis of inputs being used and levels of investment.

Afterwards, in Chap. 5, the pollutant emissions are presented and the equivalent carbon dioxide [$[(CO_2)e]$], pollutant indicator (Π_g), and ecological efficiency (ε) of hydrogen production processes are determined. Moreover, for the processes using renewable energy sources, such as in the case of ethanol, biogas, and green algae, carbon credit incorporation is also presented so as to make ecological efficiency values yet to be determined more realistic.

In Chap. 6 it is applied a multi-criteria analysis for a sustainability assessment of hydrogen production techniques in Brazil.

Finally, in Chap. 7, the main conclusions and suggestions for future studies on the matter at hand are presented.

1.3 Objectives

The present work aims to study hydrogen production processes by analyzing their technical, economic, and ecological aspects. Specific objectives are:

- Investigating the technical aspects: calculating the energy efficiency of each hydrogen production process, discussing the considerations involved and carrying out a comparative study between the determined efficiencies.
- Assessing the economic aspects: determining the hydrogen production costs per kWh of electricity produced (US$/kWh) from each process, thus aiming at a viability comparison between the considered processes.
- Investigating the ecological aspects: determining the ecological efficiency of each hydrogen production process by considering which form of hydrogen production is more environmentally correct.

References

Fatsikostas AN, Kondarides DI, Verykios XE (2002) Production of hydrogen for fuel cells by reforming of biomass-derived ethanol. Catalysis 75:145–155

Silva EP (1991) Introdução à tecnologia e economia do hidrogênio. Unicamp, Campinas, p 204. ISBN 85-268-0174-0

1.3 Objectives

The present work aims to study hydrogen production processes by analyzing their technical, economic, and ecological aspects. Specific objectives are:

- Investigating the technical aspects: calculating the energy efficiency of each hydrogen production process, discussing the considerations involved and carrying out a comparative study between the determined efficiencies.
- Assessing the economic aspects: determining the hydrogen production costs per kWh of electricity produced (US$/kWh) from each process, thus arriving at a viability comparison between the considered processes.
- Investigating the ecological aspects: determining the ecological efficiency of each hydrogen production process by considering which form of hydrogen production is more environmentally correct.

References

Fatsikostas AN, Kondarides DI, Verykios XE (2002) Production of hydrogen for fuel cells by reformation of biomass-derived ethanol. Catalysis 75:145–155
Silva EP (1991) Introdução à tecnologia e economia do hidrogênio. Unicamp, Campinas. ISBN 85-268-0177-0

Chapter 2
Hydrogen Production Processes

Lúcia Bollini Braga, Márcio Evaristo da Silva, Túlio Stefani Colombaroli, Celso Eduardo Tuna, Fernando Henrique Mayworm de Araujo, Lucas Fachini Vane, Daniel Travieso Pedroso, and José Luz Silveira

2.1 Introduction

The main hydrogen production processes can be classified into electrolysis, photolysis, and thermolysis. Electrolytic hydrogen production processes involve the use of electric or thermal energy to trigger a chemical reaction for splitting water molecules into hydrogen and oxygen. The main examples of electrolytic processes are water electrolysis (conventional process) and thermolysis (steam electrolysis). Photolytic processes involve technologies that use the energy of light, and its main examples are the photobiological and photoelectrochemical systems. Hydrogen production through thermochemical processes essentially comprise the raw material, being either from fossil or renewable sources, heat and catalysts so as to trigger chemical reactions for transforming the raw material (for example, ethanol, natural gas, methanol, gasoline) into hydrogen. The main thermochemical hydrogen production processes are: biomass gasification and pyrolysis, steam reforming, partial oxidation, autothermal and oxidative reforming (Braga 2010; Silva 2010).

L.B. Braga • M.E. da Silva • T.S. Colombaroli • C.E. Tuna • F.H.M. de Araujo • L.F. Vane • D.T. Pedroso
Group of Optimization of Energy Systems—GOSE, College of Engineering of Guaratinguetá, Institute of Bioenergy Research—IPBEN, São Paulo State University—UNESP, Dr. Ariberto Pereira da Cunha Avenue, 333, Guaratinguetá 12516-410, SP, Brazil

J.L. Silveira (✉)
Group of Optimization of Energy Systems—GOSE, College of Engineering of Guaratinguetá, Institute of Bioenergy Research—IPBEN, São Paulo State University—UNESP, Dr. Ariberto Pereira da Cunha Avenue, 333. Pedregulho, Guaratinguetá 12516-410, São Paulo, Brazil
e-mail: joseluz@feg.unesp.br

J.L. Silveira (ed.), *Sustainable Hydrogen Production Processes*, Green Energy and Technology, DOI 10.1007/978-3-319-41616-8_2

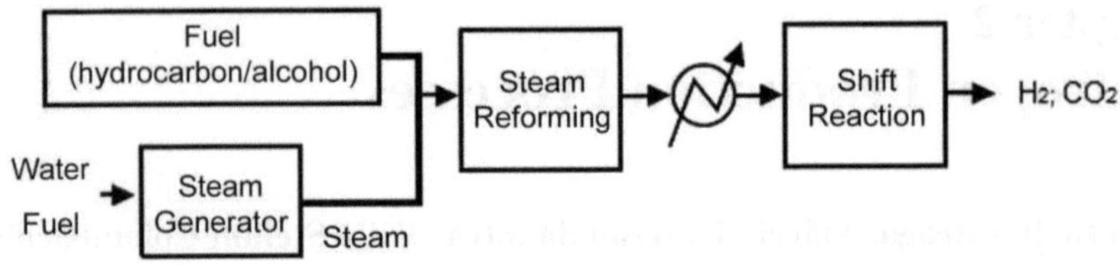

Fig. 2.1 Steam reforming process flow chart (Adapted from Spath and Mann 2000)

2.2 Steam Reforming for Hydrogen Production

Steam reforming has been used as the main process for hydrogen production, which accounts for 50 % of the world's total production. The popularity of this process derives from its high conversion efficiency and cost-effectiveness in comparison with other processes (Chena et al. 2008). Figure 2.1 shows a simple reforming process flow chart.

As illustrated in Fig. 2.1, this process occurs in two main steps, one that occurs at high temperatures (*steam reforming*), in which the fuel (hydrocarbon or alcohol) reacts with steam and is converted into a gaseous mixture of H_2, CO, CO_2, hydrocarbon or alcohol, and unreacted steam, and another one which occurs at lower temperatures in a *shift* reactor (water-gas shift reaction), where the CO present in the synthesis gas reacts with H_2O to produce additional CO_2 and H_2.

Depending on the specifications of the fuel cell being used, an additional step of CO removal is necessary to purify the synthesis gases.

The main reactions involved in the steam reforming process are shown in Eqs. (2.1)–(2.3). Many chemical reactions can occur while the steam reforming reaction takes place simultaneously, and such reactions are presented in Eqs. (2.4)–(2.7) (Trane et al. 2012).

1. *Overall Reforming Reaction*
   The overall reaction of fuel conversion into hydrogen by steam reforming is shown in Eq. (2.1).

$$\text{Overall Reaction:}\ \ \text{Fuel} + H_2O_{(g)} \leftrightarrow CO_{2(g)} + H_{2(g)} \tag{2.1}$$

2. *Steam Reforming Reaction*
   Equation (2.2) shows the steam reforming reaction which consists of an endothermic catalytic reaction of the fuel with steam, mainly producing carbon monoxide and hydrogen:

$$\text{Reforming Reaction:}\ \ \text{Fuel} + H_2O_{(g)} \leftrightarrow CO_{(g)} + H_{2(g)} \tag{2.2}$$

3. *Water-Gas Shift Reaction*
   Equation (2.3) shows the catalytic water-gas shift reaction that produces additional hydrogen and eliminates part of the carbon monoxide, which is conducted through a catalytic reactor called as the *shift reactor*, i.e., the carbon monoxide reacts with steam so as to form carbon dioxide and hydrogen. This reaction is

carried out at lower temperatures, ranging from 200 to 300 °C (Casanovas et al. 2010).

$$\text{Shift Reaction: } CO_{(g)} + H_2O_{(g)} \leftrightarrow CO_{2(g)} + H_{2(g)} \quad (2.3)$$

4. *Methane Formation Reaction*
Equation (2.4) shows one of the reactions that can occur during the reforming process. The methane formation reaction is unwanted, since part of the produced hydrogen reacts with carbon monoxide, thus decreasing its composition in the final synthesis gas.

$$\text{Methane forming reaction: } CO_{(g)} + 3H_{2(g)} \leftrightarrow CH_{4(g)} + H_2O_{(g)} \quad (2.4)$$

5. *Boudouard's Carbon Formation Reaction*
One should consider, moreover, the possibility of producing carbon by decomposing carbon monoxide through the so-called Boudouard reaction, as shown in Eq. (2.5).

$$\text{Boudouard reaction to form carbon: } 2CO_{(g)} \rightarrow CO_{2(g)} + C_{(s)} \quad (2.5)$$

In order to illustrate the steam reforming processes, the pictures of two prototypes developed by the Group of Optimization of Energy Systems are shown in Figs. 2.2 and 2.3.

Figure 2.2 shows the prototype for the steam reforming of ethanol, which was funded by the P&D ANEEL Project of CEMIG (Companhia Energética de Minas Gerais). This equipment is powered by electricity, consisting of a steam generator, a steam reforming reactor, and a *shift reactor*.

Figure 2.3 shows a prototype for the steam reforming of biogas, which is also an electric prototype, consisting of a steam generator, a steam reforming reactor, and a shift reactor, all developed through the Research in Public Policies—FAPESP.

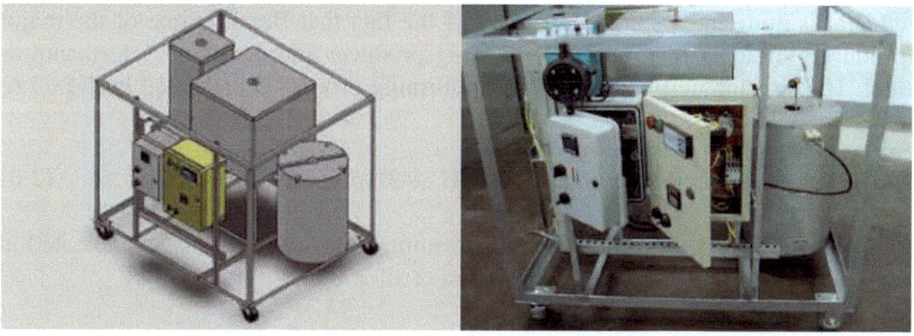

Fig. 2.2 Prototypes for the steam reforming process of ethanol

Fig. 2.3 Prototype for the steam reforming process of biogas

2.2.1 *Steam Reforming of Alcohols*

According to Silva (2005), a fairly feasible alternative is the production of hydrogen via the steam reforming of alcohols. In particular, the steam reforming of ethanol is quite interesting on account of the fact that Brazil is one of the major producers of sugarcane, being the leading producer and distributor of ethanol as fuel. The stoichiometry of the overall reforming process is presented by Eq. (2.6) (Silveira et al. 2009).

$$\text{Overall Reaction:}\ C_2H_5OH + 3H_2O \leftrightarrow 2CO_2 + 6H_2 \quad (2.6)$$

This equation is derived from the reforming and shift reactions, as indicated in Eqs. (2.7) and (2.8), respectively (Silveira et al. 2009).

$$\text{Reforming Reaction: } C_2H_5OH + H_2O \leftrightarrow 2CO + 4H_2 \quad (2.7)$$

$$\text{Shift Reaction: } 2CO + 2H_2O \leftrightarrow 2CO_2 + 2H_2 \quad (2.8)$$

The steam reforming reaction of ethanol occurs at temperatures between 400 and 700 °C (Saebea et al. 2011), and the shift reaction occurs at temperatures between 200 and 300 °C, as previously presented in Sect. 2.1.

Among the methods of hydrogen production is the steam reforming of methane, which is a process that has been intensively studied. Another alternative which is quite feasible is the production of hydrogen via the steam reforming of alcohols. In particular, the steam reforming of ethanol is rather interesting, once again being due to the fact that Brazil is one of the major producers of sugarcane, and the leading producer and distributor of ethanol as fuel. Furthermore, the overall reaction of hydrogen production from ethanol corresponds to 6 mol of hydrogen formation per mole of ethanol consumed, according to the stoichiometry in Eq. (2.9).

$$C_2H_5OH + 3H_2O \leftrightarrow 2CO_2 + 6H_2 \quad (2.9)$$

This reaction, however, occurs in two steps: one at high temperatures (*steam reforming*), in which ethanol is converted into a gaseous mixture of H_2, CO, CO_2, CH_4, and unreacted H_2O, and another at lower temperatures (water-gas shift reaction WGS), in which CO reacts with H_2O to form additional CO_2 and H_2. Due to the shift reaction being limited by the equilibrium constant, the conversion of CO is incomplete and an additional step of CO removal is required to purify the synthesis gases. The steam reforming of ethanol technology involves a catalytic process that consists of an endothermic reaction between ethanol and steam. The chemical thermodynamic analysis of the steam reforming process of ethanol for obtaining hydrogen reveals that the process does not occur in a single step, and is linked to the following reactions:

- *Overall Steam Reforming Reaction of Ethanol:*
  The overall reaction of ethanol conversion into hydrogen produces 6 mol of hydrogen from 1 mol of ethanol through the steam reforming process, as shown in Eq. (2.10) using a steam/ethanol molar ratio of 3:

$$\text{Overall Reaction: } C_2H_5OH_{(g)} + 3H_2O_{(g)} \underset{\text{Endothermic}}{\overset{\text{Catalysts}}{\rightleftarrows}} 2CO_{2(g)} + 6H_{2(g)} \quad (2.10)$$

- *Steam Reforming Reaction:*
  Equation (2.11) shows the steam reforming reaction which consists of an endothermic reaction of ethanol, mainly forming carbon monoxide and hydrogen through a steam reforming reactor with a catalytic bed at temperatures ranging between 973 and 773 K (external steam reforming: reformers) or through fuel cells, e.g., MCFC and SOFC, which are capable of performing internal steam reforming.

$$\text{Steam Reforming Reaction: } C_2H_5OH_{(g)} + H_2O_{(g)} \underset{773-923\text{ K}}{\overset{\text{NiCu-Al}_2\text{O}_3}{\rightleftarrows}} 2CO_{(g)} + 4H_{2(g)} \tag{2.11}$$

- *Water-Gas Shift Reaction:*
  Equation (2.12) shows the water-gas shift reaction that is aimed at producing hydrogen and eliminating part of the carbon monoxide, which is accomplished through a catalytic reactor called as the shift reactor, i.e., carbon monoxide reacts with steam to form carbon dioxide and hydrogen. This reaction occurs at lower temperatures, ranging from 473 to 523 K, and on account of the water-gas shift reaction being limited by the equilibrium constant, the carbon monoxide conversion is incomplete, thus requiring an additional removal step in the case of feeding a PEM fuel cell whose operation requires minimal amounts of carbon monoxide.

$$\text{Water-Gas Shift Reaction: } CO_{(g)} + H_2O_{(g)} \underset{473\ -523\text{ K}}{\overset{\frac{\text{Cu}}{\text{ZnO}}\text{-}\gamma\text{Al}_2\text{O}_3}{\rightleftarrows}} CO_{2(g)} + H_{2(g)} \tag{2.12}$$

- *Methane Formation Reaction:*
  Many chemical reactions can occur while the steam reforming reaction of ethanol is taking place simultaneously. Equation (2.13) shows the most significant reaction, i.e., the methane formation reaction (CH_4) from (CO), which should be added to the steam reforming reaction.

$$\text{Methane formation reaction: } CO_{(g)} + 3H_{2(g)} \leftrightarrow CH_{4(g)} + H_2O_{(g)} \tag{2.13}$$

- *Boudouard's Carbon Formation Reaction:*
  One should consider, moreover, the possibility of forming carbon by decomposing carbon monoxide through the so-called Boudouard reaction, as shown in Eq. (2.14).

$$\text{Boudouard carbon formation reaction: } \rightarrow 2CO_{(g)} + CH_{2(g)} + C_{(s)} \tag{2.14}$$

The equilibrium constants associated with the reactions represented by Eqs. (2.15)–(2.18) can be expressed as:

$$K_1 = \frac{y^2{}_{CO_2} y^6{}_{H_2}}{y_{EtOH} y^3{}_{H_2O}} P^4 \tag{2.15}$$

$$K_2 = \frac{y^2{}_{CO} y^4{}_{H_2}}{y^2{}_{EtOH} y_{H_2O}} P^4 \tag{2.16}$$

$$K_3 = \frac{y_{CO_2} y_{H_2}}{y_{CO} y_{H_2O}} \tag{2.17}$$

$$K_4 = \frac{y_{CH_4} y_{H_2O}}{y_{CO} y^3_{H_2}} P^{-2} \tag{2.18}$$

In which: P is the total pressure and y_i are the molar fractions of the gaseous components, which can be expressed by Eq. (2.19).

$$y_i = \frac{n_i}{n_{TOT}} \tag{2.19}$$

The equilibrium constants of reactions are related to the Gibbs free energies of the molecules involved in the equilibrium, which can be determined once the absolute temperature is specified. Once the temperature is determined, it is possible to calculate the equilibrium constants from thermodynamic data, as described by Maggio et al. (1998).

2.2.2 *Physicochemical Analysis*

2.2.2.1 Temperature Influence

The *Gibbs free energy* dependence on temperature can be expressed in several different ways, depending on the problem scale, as shown in Eqs. (2.20) and (2.21) and in Fig. 2.4.

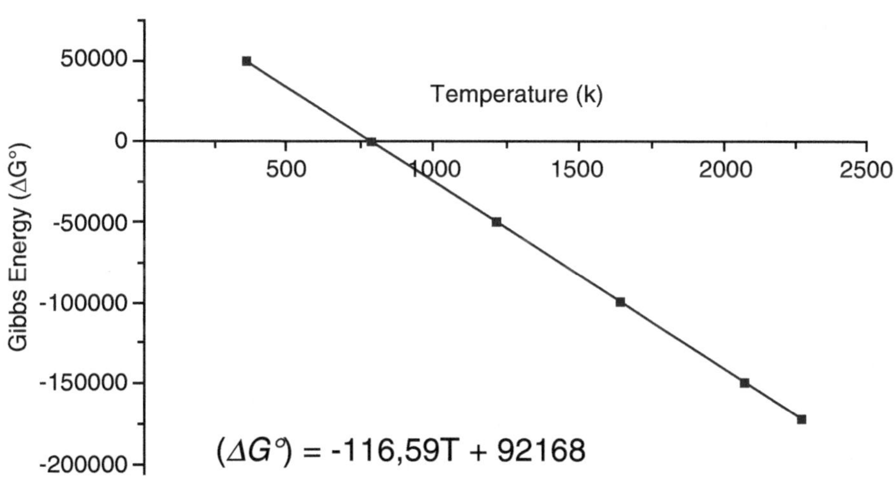

Fig. 2.4 Gibbs free energy variation as a function of temperature

$$\Delta G^0 = \Delta H^0 - T \cdot S^0 \tag{2.20}$$

$$d\left(\frac{\Delta G^{o}}{T}\right) = \frac{-\Delta H^0}{T^2} \tag{2.21}$$

Figure 2.4 shows the Gibbs energy variation as a function of temperature for the steam reforming of ethanol.

The reforming reaction is endothermic, ΔH^o is positive and the equilibrium constant increases as temperature does. Figure 2.4 shows the Gibbs free energy change (ΔG^o) as a function of temperature, in which it can be observed that it becomes null once it is close to 790 K, thus indicating that the reaction occurs at high temperatures. The straight line intersection with the horizontal axis presents a positive value, which indicates that the reaction is endothermic.

According to Le Chatelier's principle, increasing the temperature of the steam reforming reaction of ethanol will lead to a greater formation of products. Therefore, the overall reaction of the steam reforming of ethanol is favored by temperature increase, i.e., the chemical equilibrium of products is increased as temperature is raised.

2.2.2.2 Equilibrium Composition

The reaction's degree of advancement and Gibbs free energy decrease will continue until the system's Gibbs free energy reaches a minimum value, condition in which the reaction will be in equilibrium. The equilibrium composition as a function of temperature can be determined by setting the chemical equilibrium of the overall reaction, (Eq. (2.9)), as shown in Table 2.1.

By assuming an ideal behavior and disregarding the fugacity coefficients, equilibrium constant K equals Kp (equilibrium constant as a function of the partial pressure of each component). Once the molar fractions of each component in equilibrium are known, the equilibrium constant, (Eq. (2.9)), and the degree of advancement (α) of the overall steam reforming of ethanol reaction are determined, as shown in Table 2.2.

Table 2.1 Chemical equilibrium of the overall steam reforming reaction

	$C_2H_5OH +$	$3H_2O$	$\leftrightarrow 2CO_2$	$6H_2$
Start				
N° mols n_i	n	$3n$	0	0
N° mols total $n_{TOT}^{(start)}$	$n_{TOT}^{(start)} = n + 3n = 4n$			
At equilibrium				
N° mols n_i	$n(1-\alpha)$	$3n(1-\alpha)$	$2n\alpha$	$6n\alpha$
N° mols total $n_{TOT}^{(equil.)}$	$n_{TOT}^{(equil.)} = 4n(1+\alpha)$			
Molar fractions y_i	y_{EtOH}	y_{H_2O}	y_{CO_2}	y_{H_2}
	$\frac{1}{4}(1-\alpha)/(1+\alpha)$	$\frac{3}{4}(1-\alpha)/(1+\alpha)$	$\frac{1}{2}\alpha/(1+\alpha)$	$3\frac{1}{2}\alpha/(1+\alpha)$

*(i) represent the overall reaction components

Table 2.2 Equilibrium constant and degree of advancement

Equilibrium constant	K	$K = \frac{3^3 \cdot \alpha^8 \cdot P^4}{(1+\alpha)^4 \cdot (1-\alpha)^4} = \frac{27 \cdot \alpha^8}{(1-\alpha^2)^4} \cdot P^4$
Degree of advancement	α	$\alpha = \frac{\sqrt[8]{K}}{\sqrt{\sqrt[4]{K} + \pm\sqrt[4]{27 \cdot P}}}$

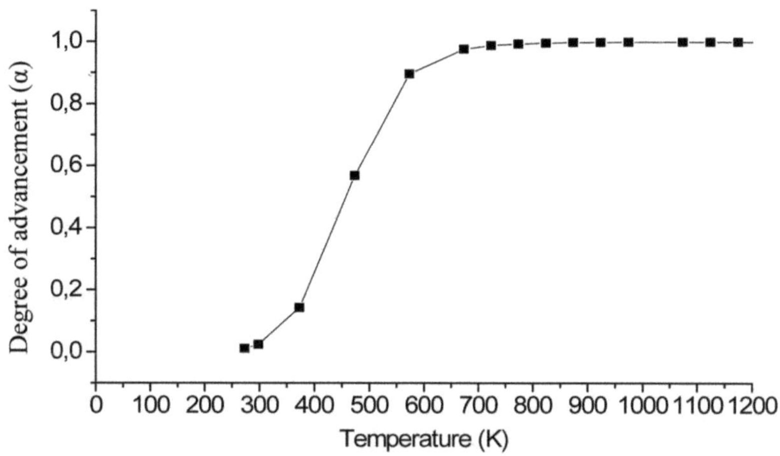

Fig. 2.5 Degree of advancement of the overall steam reforming reaction of ethanol as a function of temperature

The values of equilibrium constant, degree of advancement, molar fractions of hydrogen and ethanol as a function of temperatures ranging from 273 to 1473 K and pressure of 0.101 MPa were calculated from the equations presented in Table 2.2. The degree of advancement as a function of temperature can be analyzed through Fig. 2.5.

It can be observed that the increase in temperature favors the advancement of the overall steam reforming reaction of ethanol, consequently the hydrogen production itself. The percentages of hydrogen and carbon dioxide produced as a function of temperature, and the percentages of hydrogen produced and remaining ethanol as a function of temperature, respectively, can be analyzed in Figs. 2.6 and 2.7.

It can be observed that, in Figs. 2.6 and 2.7, hydrogen production is favored by increased temperature, reaching a maximum value of production that is close to 600–650 K, then remaining constant from this temperature range.

2.2.2.3 Pressure Influence

According to Le Chatelier's principle, an increase in the operating pressure of the overall steam reforming reaction of ethanol is going to lead to a shift in the equilibrium of the reaction in order to decrease the number of moles. That is, an

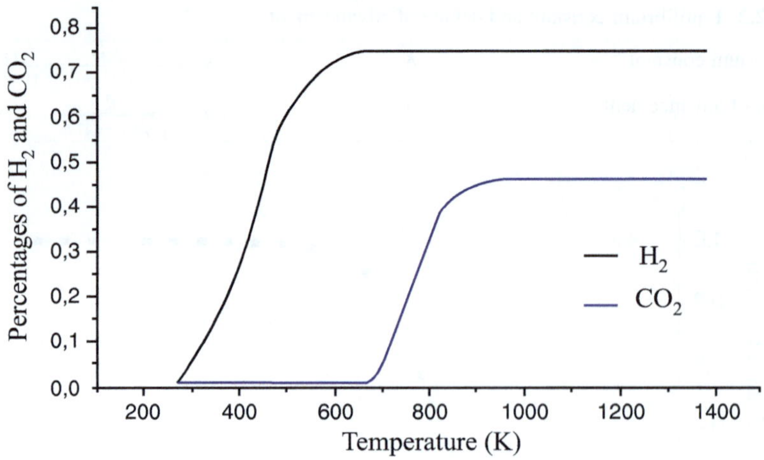

Fig. 2.6 Percentages of hydrogen and carbon dioxide produced as a function of temperature

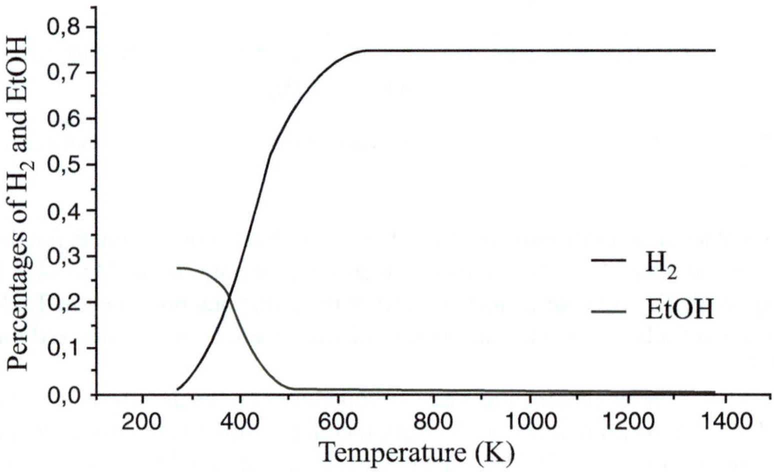

Fig. 2.7 Hydrogen production and remaining ethanol percentage as a function of temperature

increase in pressure shifts the equilibrium towards the reactants. The analysis of equations in Table 2.2 shows that increased pressure causes a decrease in the degree of advancement of the overall steam reforming reaction of ethanol and, consequently, in hydrogen production.

Figure 2.8 shows this behavior of the degree of advancement of the overall steam reforming reaction of ethanol as a function of different pressures and temperatures.

It can be observed that, with the exception of this behavior in the temperature range of 573 K, the degree of advancement does not vary significantly as pressure does. Therefore, it can be concluded that pressure does not favor hydrogen production.

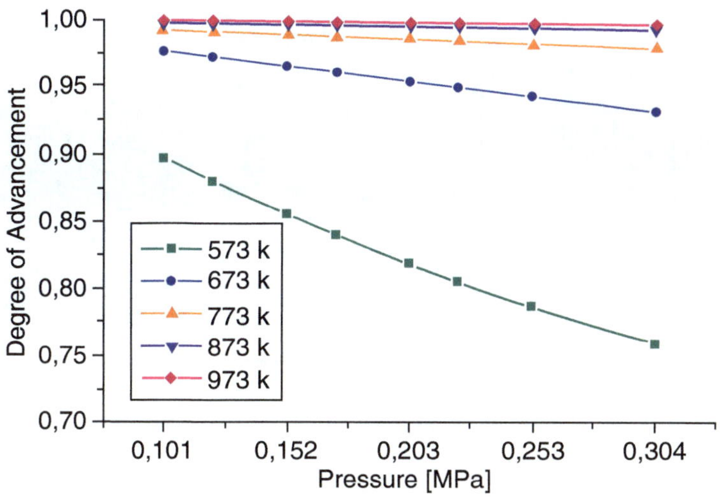

Fig. 2.8 Degree of advancement as a function of pressure at different temperatures

2.2.3 Steam Reforming of Natural Gas and Biogas

The steam reforming process using natural gas accounts for 50 % of the world's hydrogen production. Natural gas has a similar origin to petroleum: it originated over millions of years through the decomposition of plants and animals, i.e., it is a nonrenewable fuel. Its composition shows some variations according to its origin and processing, being composed mainly of methane (about 90 %), ethane (from 5 to 8 %), propane, and traces of heavier hydrocarbons, moreover, it presents inert gases such as nitrogen, carbon dioxide, and helium (Krona et al. 2012).

Biogas can be a renewable alternative as raw material for conventional steam reforming. As natural gas, it is a mixture which is composed mainly of methane, and it results from an anaerobic fermentation of organic matter. In addition, if compared to natural gas, it offers the following benefits: it reduces methane emissions by the use of organic matter that would be released into the environment (methane is 21 times more harmful than carbon dioxide as regards the greenhouse effect) and it can be produced commercially on a large scale through the decomposition of organic matter from various sources (agricultural waste, tree-pruning waste, organic waste, industrial waste, sewage, animal waste, etc.) (Kothari et al. 2008). Depending on the biodigestion technology and the raw material being used, the composition of biogas varies between 45 and 75 % of CH_4 and 25–55 % carbon dioxide, in addition to containing traces of hydrogen, sulfur, ammonia, and steam (Scholz et al. 2013).

Hydrogen production processes by the steam reforming of natural gas and biogas are based mainly on the steam reforming of methane, since it is the major constituent of both fuels. Reforming processes have contributed effectively to an increased

Fig. 2.9 SEM (Scanning Electron Microscope) images of carbon deposition on Nickel cermets (Reproduced from Takeguchi et al. 2002)

utilization of fuel cells because, through these processes, the possibilities of using fuel in these devices are increased, in addition to minimizing typical problems that arise from the use of such cells.

A major drawback found in different types of existing fuel cells is that the inputs used for feeding them should be homogeneous. Low-temperature fuel cells, such as polymer electrolyte membrane fuel cells, are only run on high-purity hydrogen; on the other hand, high-temperature cells can be run on more diverse compositions, where hydrocarbon-based fuels are mainly used because of their high hydrogen content. However, this often results in the formation of layers of solid carbon on the anode, which is a phenomenon known as carbon deposition. It drastically reduces the electrochemical performance of the fuel cell because it decreases the electrochemically active area, resulting in an increase in polarization strength and probably causing irreversible microstructural damages, such as metal dusting or catastrophic carburization, which leads to the disintegration of the metal, thus producing thin metallic dust and carbon (Kuhn and Kesler 2015). Figure 2.9 illustrates the carbon deposition on Nickel cermets,[1] being that (a) and (b) show a sample of Nickel cermet with yttrium stabilized on zirconia (Ni-YSZ) and (c) illustrates a nickel catalyst supported on zirconium II oxide (ZrO_2).

In a study conducted by Wongchanapai et al. (2013), it was shown the influence of the addition of reforming agents, like oxygen and steam, on the performance of solid oxide fuel cells. In Fig. 2.9 it is noted the influence of temperature and composition on carbon deposition on the anode. The temperature lines in red are the deposition boundaries; each point belonging to this line is the limit composition so that there is no carbon formation in the solid oxide cell. Compositions that are above the aforementioned lines favor carbon deposition, and those below the lines

[1]Cermet: A composite material composed of ceramic (CER) and metallic (MET) materials.

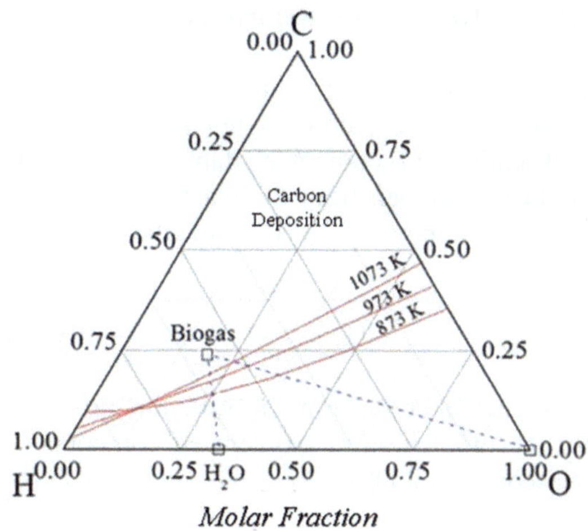

Fig. 2.10 Ternary diagram C-H-O with carbon deposition boundary in a solid oxide fuel cell (Reproduced from Wongchanapai et al. 2013)

contribute to non-deposition. It is reasonable to expect that, once reforming agents are added (H_2O_{steam} and O_2), the molar fraction will tend to be at the base of the ternary diagram (decreased concentration of C), i.e., for non-carbon deposition. Temperature also contributes to preventing carbon formation, i.e., the more it increases, the higher the deposition boundary will be (Fig. 2.10).

2.2.3.1 Partial Oxidation Reforming

- Partial oxidation:
  Partial oxidation (Eq. (2.22)) is similar to the total combustion of methane (Eq. (2.23)), but it differs due to an insufficient amount of oxygen for complete combustion, thus producing only hydrogen and carbon monoxide as by-products.

$$\text{Partialoxidation: } CH_4 + \frac{1}{2}O_2 \rightarrow CO + 2H_2 \quad \Delta H = -35.7\,\text{kJ/mol} \tag{2.22}$$

$$\text{Total combustion: } CH_4 + 2O_2 \rightarrow CO_2 + 2H_2O \quad \Delta H = -802\,\text{kJ/mol} \tag{2.23}$$

The presence of O_2 reduces carbon deposition at high temperatures, which increases the lifetime of the catalyst; however, the need for pure O_2 pushes up the costs of the plant, as it requires a cryogenic air separation unit. On an industrial scale, the process of partial oxidation of methane is not fully established, mainly because there might be a power supply and reaction medium with CH_4 and O_2, thus resulting in complete combustion and entailing risks of explosions (Vasconcelos 2006).

2.2.3.2 Dry or CO_2 Reforming

The CO_2 reforming of methane, also known as dry reforming (Eq. (2.24)), consists in an alternative route for synthesis gas production.

The high CO_2 content available on biogas offers a very interesting alternative to produce hydrogen from the reaction of CO_2 with methane, although the yield is lower if compared to that of the steam reforming reaction (Piroonlerkgul et al. 2008).

$$CH_4 + CO_2 \leftrightarrow 2CO + 2H_2 \quad \Delta H = 247\,\text{kJ/mol} \tag{2.24}$$

This is an endothermic process that produces a H_2/CO ratio of 1, being suitable for the production of oxygenated compounds and high-purity carbon monoxide (Vasconcelos 2006).

2.2.3.3 Steam Reforming

The steam reforming of methane for hydrogen production achieves high conversion efficiency, and presents itself as a simple, low-cost technology. It is probably the most common method of H_2 production in chemical industries, where the steam reforming of natural gas accounts for approximately 95 % of the hydrogen produced in the United States.

In a study conducted by Piroonlerkgul et al. (2008), it was analyzed systems using different reforming agents, such as steam, air, and a mixture of steam and air, and it was found that steam was the most suitable agent, generating greater energy density than other air-powered systems; however, it presented a slight decline in electrical efficiency due to expenditures for generating steam. Table 2.3 compares the efficiencies and costs of the major technologies for hydrogen production:

Table 2.3 Efficiencies and costs of major technologies for hydrogen production (Adapted from T-Raissi and Block 2004)

Process	Efficiency [%]	Costs associated with SRM
Steam reforming of methane (SRM)	70–80	1
Pyrolysis of methane/natural gas	72–54	0.9
Dry reforming of landfill gas	47–58	~1
Partial oxidation of heavy oils	70	1.8
Steam reforming of waste oil	75	<1
Coal gasification	60	1.4–2.6
Partial oxidation of coal	55	–
Water electrolysis using wind power	10	>3
Water electrolysis at high temperatures	48	2.2
Thermochemical water splitting	35–45	6
Biomass gasification	45–50	2.0–2.4

The reforming process occurs by feeding the system with biogas (syngas, natural gas, or pure methane), and then the superheated steam passes through a bank of tubes on a fixed-bed reactor with nickel as catalyst, thus generating H_2 and CO, according to Eq. (2.25):

$$CH_4 + H_2O_{(v)} \rightarrow CO + 3H_2 \quad \Delta H = 206\,\text{kJ/mol} \tag{2.25}$$

The thermodynamic equilibrium of this reaction depends on temperature, pressure, and steam-to-carbon molar ratio conditions, usually referred to as S/C ratio. The steam reforming reaction is strongly endothermic and favored by high temperature and pressure conditions. The reforming rates are controlled by the reaction kinetics, the gas's mass transfer rate to the surface of the catalyst, the diffusion rate through the pores of the catalyst, and the transfer of heat through the tubes of the reformer (Alves 2005).

Afterwards, the steam reforming products are directed to shift reactors, whose function is to reuse reagents that were not used in the steam reforming process, and the products of partial reactions that occurred in the reformer (such as CO), producing additional H_2, as shown in Eq. (2.26).

Shift reaction:

$$CO + H_2O \rightarrow CO_2 + H_2 \quad \Delta H = -41\,\text{kJ/mol} \tag{2.26}$$

Carbon monoxide (CO), even in small concentrations, can be extremely harmful in certain fuel cell types. The overall steam reforming reaction is represented in Eq. (2.27):

Overall reaction:

$$CH_4 + 2H_2O \rightarrow CO_2 + 4H_2 \quad \Delta H = 165\,\text{kJ/mol} \tag{2.27}$$

2.2.3.4 Autothermal Reforming

Autothermal reforming is a combination of steam reforming and partial oxidation. It has aroused interest on account of removing the disadvantages of other reactions: low energy requirement to compensate for the effects of endothermic reactions (steam reforming) and exothermic reactions (partial oxidation). In addition, it presents low specific consumption, and there is effective control over the H_2/CO ratio by adjusting the feed rates of oxygen and steam (Cai et al. 2006).

There is great interest in this type of reforming process, especially for hydrogen production in small-scale applications, such as distributed generation and small stationary stations, due to its high efficiency, fast startup, and response (Araki et al. 2010). Equation (2.28) shows a typical autothermal reforming reaction.

$$CH_4 + \tfrac{1}{2}O_2 + H_2O \rightarrow CO_2 + 3H_2 \tag{2.28}$$

2.2.4 *Catalyst for Steam Reforming Reactions*

As the steam reforming process is composed of catalytic reactions, the study of catalysts becomes an essential part in this chapter. The use of suitable catalysts can minimize unwanted reactions, such as Boudouard's reaction (Duane et al. 2002).

The greatest difficulty of reforming reactions to produce hydrogen and synthesis gas lies in obtaining stable catalysts, which are simultaneously selective for CO_2 and H_2 and resistant to metallic synthesis and coke deposition (Abreu et al. 2012). Two ways to minimize the formation of coke on the catalyst are highlighted as follows: by increasing the steam-to-carbon (S/C) molar ratio (Wang et al. 2004a, b) and the selection of a suitable medium, as well as its metallic phase and preparation method (Silva 2010).

2.2.4.1 Catalyst for the Steam Reforming of Ethanol

For this type of reforming process, two catalysts available in literature were selected, which work at different temperatures and have different conversion rates, which are presented in Table 2.4.

The catalyst 5 %Ni-5 %Cu/γ-Al_2O_3 offers advantages in comparison with 20 % Ni/γ-Al_2O_3 because it presents the best conversion rate and works at a lower temperature. Therefore, this catalyst was considered as the basis for the study of this kind of reforming process. It was also used in the prototype for the steam reforming of ethanol, developed by the Group of Optimization of Energy Systems.

2.2.4.2 Catalyst for the Steam Reforming of Natural Gas

For this kind of reforming process, two catalysts available in literature were also selected, which work at different temperature ranges and present different methane conversion rates as indicated in Table 2.5.

Table 2.4 Catalysts for the system of steam reforming of ethanol (Adapted from Maia et al. 2007 and Liguras et al. 2003)

Catalyst/support	Reaction temperature (°C)	H_2O/C_2H_6O	Ethanol conversion (%)
5 %Ni-5 %Cu/γ-Al_2O_3	400	5	97.62
20 %Ni/γ-Al_2O_3	700	3	77

Table 2.5 Catalysts for the system of steam reforming of natural gas (Adapted from Souza 2005 and Beurden 2004)

Catalyst/support	Reaction temperature (°C)	H_2O/CH_4	Methane conversion (%)
15 %Ni/ZrO_2-γ-Al_2O_3	650	1	66
2 %Ru/α-Al_2O_3	800	2.5	85

2 %Ru/α-Al_2O_3 achieves the best conversion rate, but it works at higher temperatures. According to literature, this high temperature condition is a typical feature of this kind of reforming process (Stein et al. 2009), thus it was opted for this catalyst as the basis for the study of the present reforming process.

2.2.4.3 Catalyst for the Steam Reforming of Biogas

For this type of reforming process, two types of catalysts were also selected, which are shown in Table 2.6 with their respective operating temperature and methane conversion rate.

5 % Ru/γ-Al_2O_3 achieves the lowest conversion rate and works at a lower temperature. However, it was opted for this type of catalyst as the basis for the study of this reforming process because it is similar to the one chosen for the steam reforming of natural gas and, moreover, on account of having been used in the prototype built by the Group of Optimization of Energy Systems.

According to Avraam et al. (2010), it was observed that for CH_4/CO_2 molar ratios greater than 1.5 and H_2O/CH_4 of 2, the conversion rate of methane decreased slowly, becoming practically constant at 80 %.

2.2.4.4 Shift Reaction

The shift reaction, otherwise known as *water-gas shift reaction*, is the same in all reforming processes and occurs in a second reactor. It is an industrial technology in which steam (H_2O) reacts with carbon monoxide (CO) to produce hydrogen (H_2) and carbon dioxide (CO_2), as presented in Eq. (2.26) (Haryanto, et al. 2011).

There are studies on catalysts for this process, but the most common one is shown in Table 2.7.

Thus, according to Table 2.7, it was opted for this catalyst as the basis for the study of the *shift* reaction. This was also the catalyst used in both shift reactor prototypes developed by the group of optimization of energy systems.

Table 2.6 Catalysts for the system of steam reforming of biogas (Adapted from Aizquierdo et al. 2012 and Avraam et al. 2010)

Catalyst/support	Reaction temperature (°C)	CH_4/CO_2 and H_2O/CH_4	Biogas conversion (%)
11.4 %Ni/Zr-γ-Al_2O_3	800	1.5 and 2	98.8
5 %Ru/γ-Al_2O_3	700	1.5 and 2	90.0

Table 2.7 Catalyst for the water-gas shift reaction system (Adapted from Brenna 2010)

Catalyst/support	Reaction temperature	H_2O/CO	Carbon monoxide conversion
Cu/ZnO/γ-Al_2O_3	250 °C	2	93 %

2.2.5 *Experimental Analysis of the Steam Reforming of Ethanol*

Considering the need to seek alternative energy sources, the potential of heterogeneous catalysis and the benefits obtained by the use of hydrogen, it was defined a methodology adopted for the experimental analysis of the steam reforming process proposed in this chapter, which consists in the investigation of technical aspects, such as preparation of mono- and bimetallic catalysts that are suitable for the steam reforming of ethanol, development of a method to characterize supports and catalysts, and, furthermore, conducting experimental tests on reforming prototypes with the capacity to produce 1 Nm^3/h of hydrogen.

2.2.5.1 Pure and Mixed Supports, and Mono- and Bimetallic Catalysts Preparation

According to the characteristics of each material and behavior associated with the steam reforming reactions of ethanol, it was defined the composition and preparation methods of supports and catalytic series, which are composed of mono- and bimetallic catalysts (15 %Ni, 15 %Cu, 0.5 %Pt, 0.5 %Pt-15 %Ni). Figure 2.11 shows the sequence of experimental steps used in the preparation of catalytic materials.

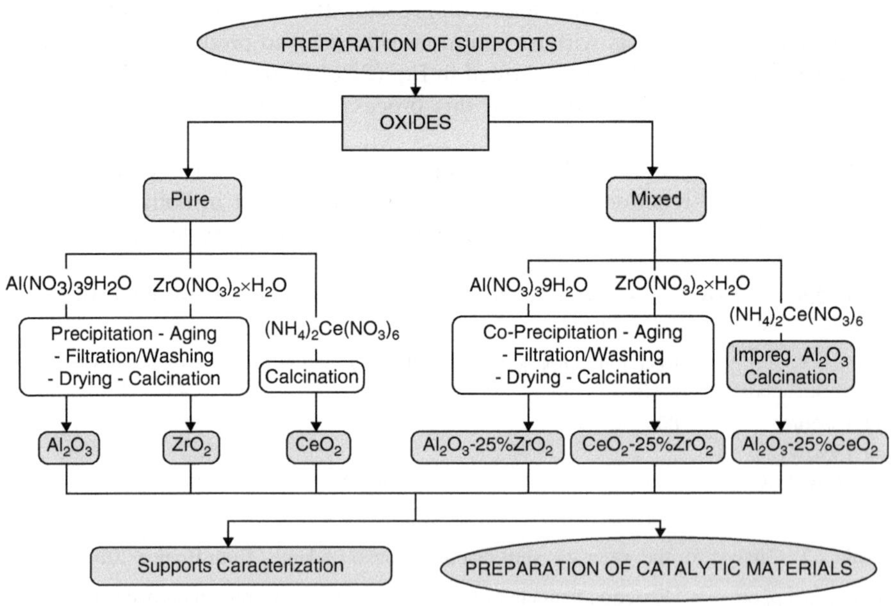

Fig. 2.11 Methods used in the preparation of catalytic materials

For each prepared support (γ-Al_2O_3; ZrO_2; CeO_2; γ-Al_2O_3 – 25 %ZrO_2; γ-Al_2O_3 – 25 %CeO_2; CeO_2 – 25 %ZrO_2), the precursor material and experimental method were employed that, according to the chemical composition of the oxide support and the characteristics observed by other researchers, would have more suitable properties to be used in the steam reforming of ethanol reactions. It can be observed in Fig. 2.11 that, in the preparation of pure oxides, experimental techniques involving precipitation, aging, filtration, washing, and heat treatments of drying and calcination were used. As for the preparation of the mixed oxides, in addition to the aforementioned techniques, the methods of coprecipitation and impregnation were used.

Materials and Reagents

The mono- and bimetallic catalysts containing 15 %Ni, 15 %Cu, 0.5 %Pt, 0.5 % Pt-15 %Ni, supported on alumina (γ-Al_2O_3), zirconium(II)oxide (ZrO_2), ceria (CeO_2), alumina-zirconia (γ-Al_2O_3-25 %ZrO_2), alumina-ceria (γ-Al_2O_3-25 % CeO_2), and ceria-zirconia (CeO_2-25 % ZrO_2) were prepared by the dry impregnation method (also called incipient impregnation) and submitted to heat treatments of washing, drying, and calcination (Table 2.8).

Preparation of Catalytic Supports

Methods of Preparation: Pure Oxides

The procedures for preparing the catalytic supports, which are pure aluminum oxide, Zirconium dioxide, and ceria oxide, are described below with the different

Table 2.8 Materials and reagents

$Al(NO_3)_3 \cdot 9H_2O$	P.A. (98 %)	Sigma-Aldrich
$ZrO(NO_3)_2$	P.A. (99 %)	Sigma-Aldrich
$Zr(OH)_4$	P.A. (97 %)	Sigma-Aldrich
$(NH_4)_2Ce(NO_3)_6$	P.A. ACS (98.5 %)	Sigma-Aldrich
$Cu(NO_3)_2 \cdot 3H_2O$	P.A. (99.999 %)	Sigma-Aldrich
$Ni(NO_3) \cdot 6H_2O$	P.A. ($\geq$97 %)	Sigma-Aldrich
$H_2PtCl_6 \cdot xH_2O$	P.A. (99.9 %—38 %Pt)	Sigma-Aldrich
NH_4OH	P.A. (28–30 % NH_3)	Sigma-Aldrich
C_2H_5OH	P.A. ACS ($\geq$99.5 %)	Sigma-Aldrich
HNO_3	P.A. ACS (63 %)	Sigma-Aldrich
HCl	P.A. ACS	Sigma-Aldrich
Argon 5.0 Analytical	(99.999 %)	White Martins
Helium 5.0 Analytical	(99.999 %)	White Martins
Ultra pure nitrogen		White Martins
Ultra pure hydrogen		White Martins
Primary standard mixtures		White Martins
H_2O		Distilled

heat treatments to which they were subjected, as well as a list of the different solids obtained with their respective nominal compositions and designations.

The prepared supports were:

- Aluminum oxide or alumina (Al_2O_3)
- Zirconium dioxide or zirconia (ZrO_2)
- Cerium oxide or ceria (CeO_2)

- **γ-Al_2O_3**

  The support γ-Al_2O_3 was prepared by precipitation using solutions of aluminum nitrate $Al(NO_3)_3{\cdot}9H_2O$ (0.5 mol/L) and ammonium hydroxide NH_4OH (6 mol/L). The precipitation consisted in dripping the aluminum nitrate solution (0.5 mol/L) evenly over the ammonium hydroxide solution (6 mol/L) with a peristaltic pump under constant stirring at $pH \geq 10$, as shown in Fig. 2.12. Afterwards, the system became idle with the purpose of aging for 16 h at 298 K. After the aging process, the precipitate was filtered and washed in a Büchner funnel with distilled water at 333 K using the system presented in Fig. 2.13a for removing the precipitating agent until the washing water reached neutral pH. Afterwards, the obtained aluminum hydroxide was dried in a Heraeus Vacutherm vacuum drying oven for 16 h at a temperature of 343 K, Fig. 2.13b. After being cooled in a desiccator, the material was ground and submitted to calcination in a Quimis muffle furnace at 773 K for 3 h, at a heating rate of 10 °C/min, Fig. 2.13c.

Fig. 2.12 Precipitation system

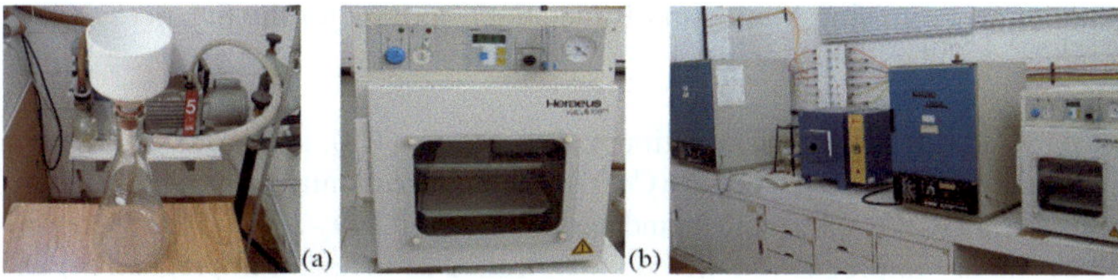

(a) (b) (c)

Fig. 2.13 Equipaments utilized: (**a**) Washing-filtration, (**b**) vacuum drying, and (**c**) calcination systems

- **ZrO_2**
  The support ZrO_2 was prepared by precipitation using solutions of zirconium oxynitrate ($ZrO(NO_{3)}\cdot\times H_{2O}$) (0.2 mol/L) and ammonium hydroxide ($NH_{4O}H$) (6 mol/L). The experimental procedures of precipitation, aging, filtering and drying, and calcination were similar to those described in the preparation of γ-alumina.
- **CeO_2**
  As for the preparation of the support $CeO_{2,}$ it was obtained by calcining the precursor Ceric ammonium nitrate $(NH_{4)2C}e(NO_{3)6}$ in a muffle furnace (Quimis) at 1073 K for 3 h, with heating control once room temperature was reached, at a rate of 10 ° C/min. In this case, the heat treatment is responsible for decomposing the precursor salt for oxide phase determination.

Methods of Preparation: Mixed Oxides

- **Al_2O_3-25 %ZrO_2**
  The mixed support γ-Al_2O_3-25 %ZrO_2 was obtained by coprecipitation with zirconium oxynitrate $ZrO(NO_{3)}\cdot\times H_{2O}$ (0.2 mol/L), ammonium nitrate nonahydrate $Al(NO_{3)3}\cdot 9H_2O$ (0.5 mol/L), and ammonium hydroxide $NH_{4O}H$ (6 mol/L) solutions which were mixed and dripped over the ammonium hydroxide solution, according to the procedures used for obtaining γ-Al_2O_3 and ZrO_2, in which the coprecipitation, aging, filtering, drying, and calcination occurred.
- **Al_2O_3-25 %CeO_2**
  The mixed oxide Al_2O_3-25 %CeO_2 was obtained by dry impregnation of alumina $Al_2O_{3,}$ using an aqueous solution of Ceric ammonium nitrate $(NH_{4)2C}e$ $(NO_{3)6}$ as precursor, so as to obtain 25 % CeO_2 content (m/m).
- **CeO_2-25 %ZrO_2**
  For the preparation of the mixed oxide CeO_2-25 %ZrO_2, it was used, in addition to the precursor Ceric ammonium nitrate solution $(NH_4)_2Ce(NO_3)_6$ (0.01 mol/L), a solution of $ZrO(NO_3)_2\cdot\times H_2O$ (0.2 mol/L). The experimental procedure for preparing the support CeO_2-25 %ZrO_2 was the same as that used for obtaining the mixed oxide γ-Al_2O_3-25 %ZrO_2, in which the coprecipitation, aging, filtering, drying, and calcination occurred.

Preparation of Promising Mono- and Bimetallic Catalysts for the Steam Reforming of Ethanol

The catalysts with nickel (Ni), platinum (Pt), copper (Cu), supported on alumina (Al_2O_3), zirconia (ZrO_2), ceria (CeO_2), alumina-zirconia (Al_2O_3-25 %ZrO_2), alumina-ceria (Al_2O_3-25 %CeO_2), and ceria-zirconia (CeO_2-25 % ZrO_2) were prepared by the dry impregnation method (also called as incipient impregnation) and submitted to washing and the heat treatments of drying and calcination, as shown in Fig. 2.14.

The supports were impregnated using aqueous solutions of nickel(II)nitrate hexahydrate $Ni(NO_3)_2{\cdot}6H_2O$, chloroplatinic acid hydrate $H_2PtCl_6{\cdot}{\times}H_2O$, and copper(II)nitrate trihydrate $Cu(NO_3)_2{\cdot}3H_2O$ as precursors in order to obtain mono- and bimetallic catalysts with the following contents: 15 % nickel (m/m), 0.5 % platinum (m/m), 15 % copper (m/m), and 15 % nickel-platinum Ni-0.5%Pt(m/m).

The volume of the impregnating aqueous solution is equal to the total porous volume of the support, i.e., it corresponds to the mass of the one that is going to be impregnated ($V_{\mathrm{solution}} = V_{\mathrm{total\ porous}} = P_{\mathrm{v}} \times \mathrm{mass}$).

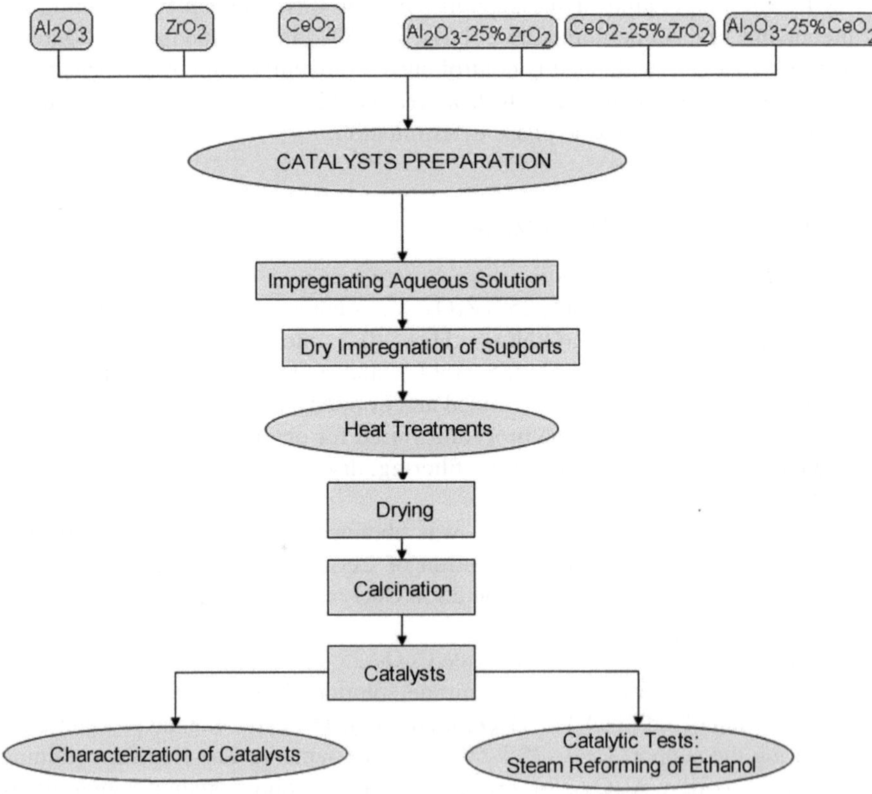

Fig. 2.14 Catalysts preparation method

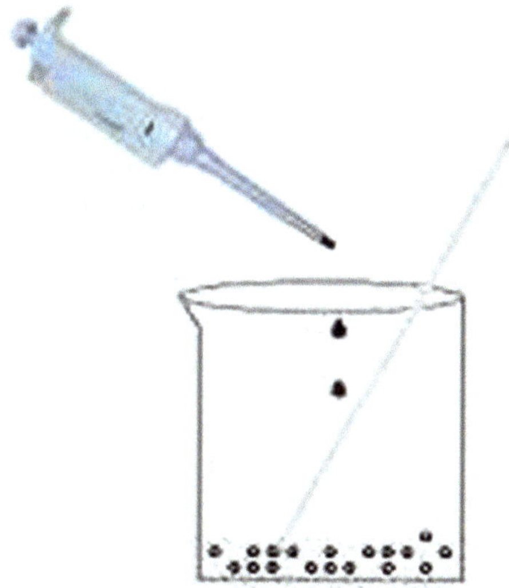

Fig. 2.15 Illustration of the system used for the dew point determination

Specific porous volumes (P_v) were determined by the dew point method, as shown in Fig. 2.15. This method consists in determining the absorption capacity of a liquid by moisturizing the support in the form of powder until it becomes wet, i.e., before clusters or a paste is formed.

Distilled water was used as liquid, which was inserted in a 10 mL automatic pipette. The water was dripped slowly and mixed with the support with a glass rod in a beaker until the dew point could be visualized. The support mass was 1 g, and at least three measurements of the absorbed liquid volume were made. Thus, the average value obtained was considered as the specific porous volume.

After impregnation, the catalysts were dried in a vacuum oven for 16 h at a temperature of 343 K (Fig. 2.13b), and then submitted to calcination that was conducted with heating control once room temperature was reached at a rate of 10 °C per minute (10 °C/min) until reaching 773 K. The calcination was performed in a muffle furnace (Quimis) for 3 h at 773 K (Fig. 2.13c) for removing water and heat stabilization.

2.2.6 *Methods for the Characterization of Supports and Catalysts*

The characterization of a catalyst results in three distinct, but interrelated pieces of information, which are: composition or chemical structure, textural properties and catalytic behavior. In principle, one should consider that almost any method of analysis related to materials science has enough potential to be used for catalyst characterization. However, experience has revealed that only a relatively restricted number of techniques are fundamentally important to catalysis sciences.

Table 2.9 Most commonly used methods to characterize catalysts

Proprieties	Characterization methods
Porosity, porous volume	Pycnometry, liquid absorption
Specific surface area	BET method
Metal specific surface area	Gas volumetry (H_2, CO)
Support crystallinity	XRD
Elemental composition	EDX
Chemical composition	AAS
Activity, selectivity	Kinetic analysis

Nevertheless, the field is extremely extensive and there are situations where there is more than just one technical alternative available to assess a particular property. The most commonly used methods that are already relatively standardized are summarized in Table 2.9.

The various characterization techniques employed for studying the prepared catalytic materials were selected from the need to better understand their physico-chemical properties and to analyze some of the most commonly used techniques in the study of the species present in a catalyst, as listed:

- Specific surface area—(BET method)
- Atomic Absorption Spectrometry—(AAS)
- Energy-dispersive X-ray spectroscopy (EDX or EDS)
- X-ray diffraction (XRD)
- Catalytic Tests involving the cyclohexane dehydrogenation Reaction

2.2.7 Considerations on the Characterization of Catalytic Materials

Therefore, for characterizing the prepared catalytic supports, it should be conducted conduct specific surface area analyses (BET), a crystallinity analysis by X-ray diffraction (XRD), and a metal content analysis of the mixed oxides through energy-dispersive X-ray spectroscopy (EDS). As regards the catalysts characterization, it should be carried out an elemental analysis by atomic absorption spectrometry, Energy-dispersive X-ray spectroscopy (EDS), Atomic Absorption Spectrometry (AAS), crystallinity analysis by X-ray diffraction (XRD), and assessments of a model reaction, such as a catalytic test of the cyclohexane dehydrogenation reaction.

The catalysts activity is generally assessed by bench tests, i.e., bed quartz microreactors at atmospheric pressure are used, which are set according to the characteristics of the reaction system adopted for the catalytic tests of the series of catalysts. The products of the heterogeneous steam reforming reaction are identified and quantified by gas chromatography.

The characterization of the prepared catalysts will, in future works, contribute to the investigation of the most promising series for applications in ethanol steam reformers by a comparison with the behavior of 6 %Ni-6 %Cu supported on commercial alumina (γ-Al_2O_3—Oxiteno, 200 m^2 g^{-1}; 0.03 m^3 g^{-1}; pellets) to be used in experimental trials carried out with the prototype for the steam reforming of ethanol.

However, from the experimental results obtained with the reforming prototype, economic and ecological analyses of the developed reforming prototype will be conducted so as to determine the cost-benefit relations of the proposed steam reforming system, and the ecological efficiency of hydrogen production generated by the reformer.

2.2.8 Prototype for the Steam Reforming of Ethanol

Reformers are devices that process ethanol into a synthesis gas which is rich in hydrogen. After that, the syngas is subjected to a purification step to meet the requirements of its application. The main components of the ethanol steam reformer are: a metering pump, a water vaporizer, a steam reforming reactor, and a water-gas shift reactor. The metering pump feeds the liquid mixture (water and ethanol) and the vaporizer, in turn, gasifies the fuel mixture (anhydrous ethanol) and water to feed the first water-gas shift reactor, i.e., the reformer, which is a stage that occurs at high temperatures where the steam reforming reactions take place in order to form a hydrogen-rich gas mixture. Then, this gaseous mixture is subjected to a catalytic process of the shift reactor that removes part of the CO and produces additional hydrogen for the synthesis gas of the process. Figures 2.16 and 2.17 show the prototypes I and II, respectively, for the steam reforming of ethanol.

It can be observed in Fig. 2.16 a scheme depicting the components of the system of hydrogen production via the steam reforming of ethanol, followed by a purification system of the syngas operating with two columns in PSA cycles (Pressure Swing Adsorption). The hydrogen-rich syngas is subjected to a system of

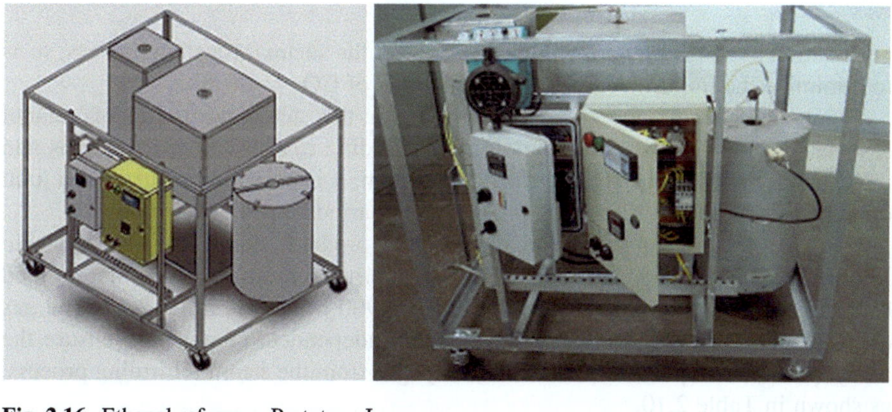

Fig. 2.16 Ethanol reformer: Prototype I

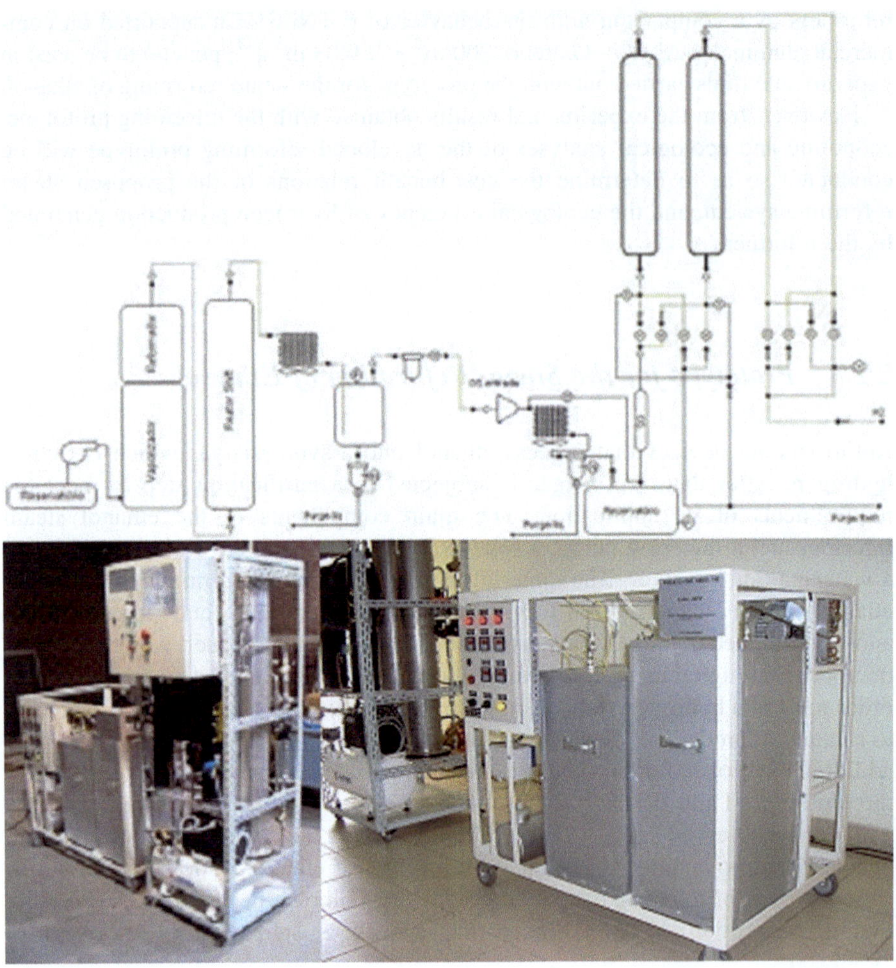

Fig. 2.17 Ethanol reformer: Prototype II

purification by molecular adsorption (PSA) after the steam reforming process, so as to minimize the impurities, mainly the amounts of CO.

Figure 2.18 shows the proposed system for the steam reforming of ethanol associated with the fuel cell technology. According to fuel cell technologies and the characteristics of the steam reforming system, it is possible to carry out a joint operation of the steam reforming process with various types of fuel cells.

The qualification and quantification of the products derived from the steam reforming of ethanol process by using the reforming prototype were determined by gas chromatography using the Varian CP-4900 Micro-GC, Fig. 2.19. The gas chromatograph was set to operate with three independent channels to ensure the analysis of the components present in the syngas from the steam reforming process, as shown in Table 2.10.

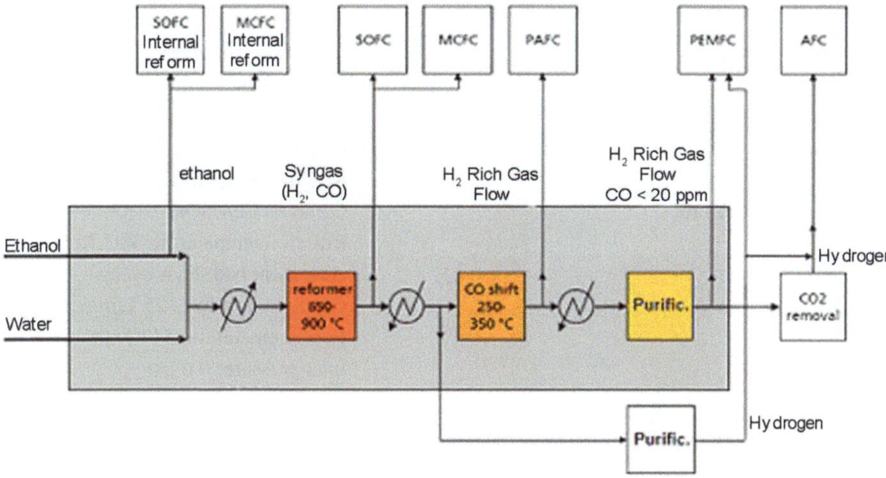

Fig. 2.18 System of steam reforming of ethanol and fuel cell technologies (Adapted from Aiche 2005)

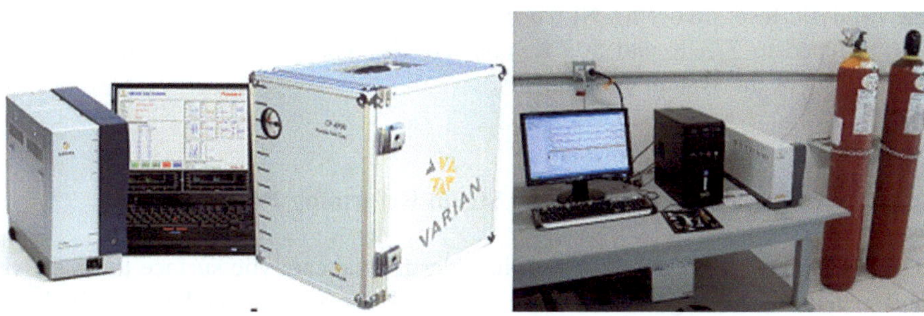

Fig. 2.19 CP-4900 Micro-GC for gas analyses

Table 2.10 Chromatograph settings

Micro GC CP-4900 settings			
Hardware	Channel 1	Channel 2	Channel 3
Carrier gas	Helium	Argon	Argon
Injector	Heated	Heated	Heated
Column	PPU	Cp sil 5	Molesieve 5A
Separation	Ar; CH_4; C_2H_4; H_2O; H_2S	Alcohols and aldehydes	H_2; Ar; O_2; N_2; CO; CH_4
Backflush	NO	NO	YES
Eliminated components	N/a	N/a	CO_2; H_2O; Hydrocarb.

Table 2.11 Operation parameters of the system

System component	Operation parameter
Vaporizer	Water/ethanol volume ratio: (1–0.9)
	Generated gas temperature: 923 K
Reformer	Catalyst: 6 %Ni-6 %Cu/γ-Al_2O_3
	Reaction temperature: 923 K; 1 atm
Water-gas shift reactor	Catalyst: Cu/ZnO/γ-Al_2O_3
	Reaction temperature: 493 K; 1 atm
Purificator	Adsorbent bed: PSA cycles
	Gas feed flow rate: 25 L/min, 0.4 bar
	Feed temperature: 313 K (max)
	Inlet pressure: 0.6 bar
	Working pressure: 6.5 bar

2.2.8.1 Experimental Trials with the Steam Reforming of Ethanol Prototype

Experimental trials were conducted with the reforming prototype. However, before the catalytic tests involving Ni-Cu/Al_2O_3 and Cu-ZnAl_2O_3, their H_2/N_2 flow rate was reduced to 773 K and 503 K, respectively. Table 2.11 shows the operation parameters of the system components.

2.2.8.2 Catalyst Activation in the Steam Reforming Prototype

The catalysts were reduced, in situ, in order for the catalytic surface to be active, and so that it is no longer oxidated on account of air exposure, and especially due to the high temperatures reached during the calcination period. Figure 2.20 shows the steam reformer during the experimental trial of catalyst activation.

Initially, the purification of the steam reforming reactors was performed with an inert gas, e.g., argon, until all oxygen was removed. The inert gas flow through the reactor was set to a given speed in order to ensure good flow distribution in the catalytic beds. After purification, the catalytic beds heating started under inert gas flow at a rate of about 323 K/h until the temperature of 423 K was reached. Afterwards, the inert gas flow was increased and the catalysts were heated to a bed top temperature of 443 K.

After a temperature of 573 K was reached in the water-gas shift reactor, about 0.5 % hydrogen was added to the inert gas flow, but not exceeding 1.0 vol.% so that no point of the bed exceeds the temperature of 723 K in the reformer, and 503 K in the water-gas shift reactor.

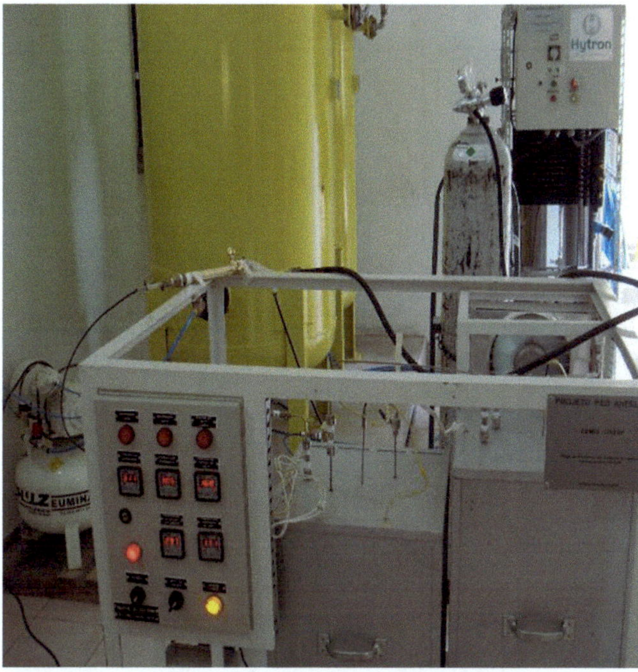

Fig. 2.20 Steam reforming of ethanol prototype: catalyst reduction

Table 2.12 Synthesis gas analysis result—before and after purification

Samples	H_2 (%mol/mol)	CO (%mol/mol)	CH_4 (%mol/mol)	CO_2 (%mol/mol)
Synthesis gas—reformer	75.24	8.64	6.79	8.12
Synthesis gas—reformer/purificator	99.84	<1.0	<1.0	<1.0

2.2.8.3 Ethanol Reformer: Prototype

Table 2.12 shows the results of the chromatographic analysis of the syngas obtained in the experimental trials of the steam reforming of ethanol system, reforming prototype II. It can be observed a high yield of hydrogen production, and low concentrations of CO, CH_4, and CO_2, which suggests that the reforming and water-gas shift reactions were strongly favored by temperature conditions of 923 K and 493 K, respectively.

Moreover, the results of the synthesis gas, collected after the purification system by molecular adsorption PSA (Pressure Swing Adsorption), show a hydrogen-rich gas flow, i.e., 99.84 % mol/mol of H_2 and low CO, CH_4, and CO_2 mol/mol percentages. The hydrogen-rich syngas flow obtained after the purification system allows using the produced hydrogen as input to generate electricity in PEM fuel cells.

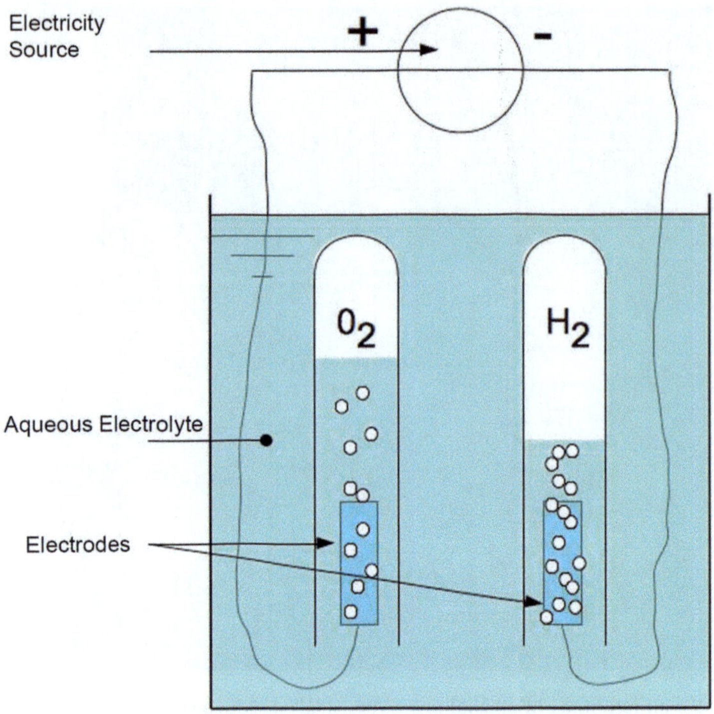

Fig. 2.21 Electrolysis process scheme (Reproduced from Hy Generation 2014)

2.3 Hydrogen Production by Water Electrolysis

Figure 2.21 illustrates a simple scheme of an electrolysis process.

The electrolysis process consists, essentially, of an electricity source (direct current), electrodes (anode and cathode), and a conductive electrolyte. Both in acid and basic electrolytes, oxidation occurs in the anode and reduction in the cathode, with a subsequent hydrogen production. The difference is in the species involved in the oxidoreduction process: on the one hand, protons (H^+) are involved, and on the other one, the hydroxyl ions (OH^-).

There are a few hindrances to the electrolysis process. One of them is the fact that a large amount of thermal energy is required to split the water molecule. It is estimated that the amount of energy required to perform the electrolysis is the same as that provided by hydrogen production. Therefore, by considering the dissipation of energy, it takes more energy to perform the electrolysis than the amount that it can generate (Lopez 2004). However, with hydrogen being used as energy input, the process becomes feasible, as it uses renewable sources for generating the energy required for the process.

The positive and negative electrodes are separated by a microporous diaphragm, which currently replaces the previously used asbestos diaphragms (Sorensen 2005).

The hydrogen ions are transported through the electrolyte due to the difference of electric potential. The alkaline component function is to improve the low ionic conductivity of water. However, this process is limited by temperatures below 100 °C in order to avoid a significant increase in the alkaline corrosion of the components (Sorensen 2005).

2.3.1 *Electrolyzers*

Conventional electrolyzers use an alkaline electrolyte solution as ionic conductor (aqueous potassium hydroxide). The electrodes are made of conventional materials, such as steel and carbon, and their anode surface is protected by traditional nickel plating (matte nickel), in order to avoid corrosion. Their operating temperature ranges from 70 to 80 °C, yielding between 70 and 80 % (Basso et al. 2013).

There are two forms of arranging electrodes in a conventional cell, which are the unipolar ones (tank type electrolyzers) and bipolar ones (filter-press electrolyzers), as shown in Figs. 2.22 and 2.23.

As for the unipolar electrolyzer, the conduction is made with the electrodes being arranged parallelly while, in the case of the bipolar electrolyzer, they are arranged serially (with the exception of those on the extremities) with one electrode working as anode in a cell and another one as cathode in the subsequent one. In the case of the unipolar electrolyzer, the electrolyte is ordinary, while it is individual in the bipolar electrolyzer (Silva 1991).

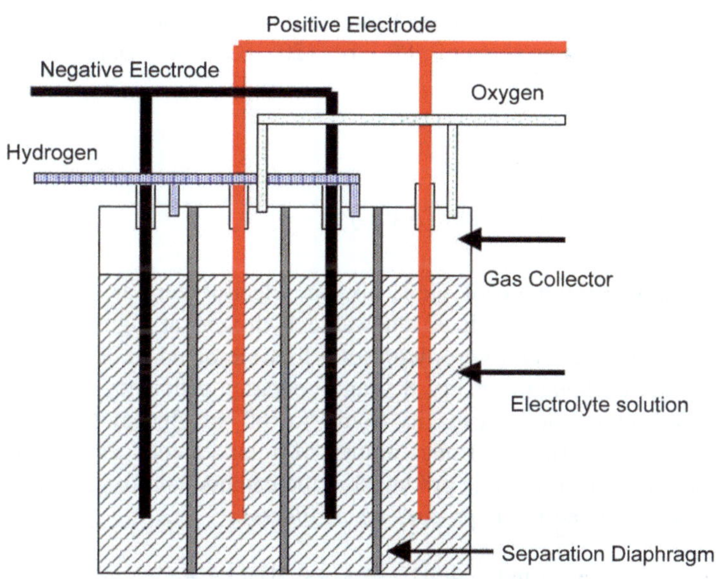

Fig. 2.22 Unipolar electrolyzer (Adapted from Kroposki et al. 2006)

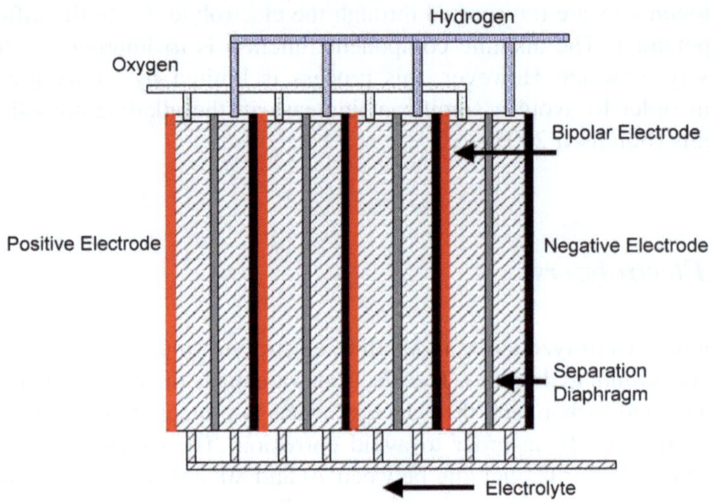

Fig. 2.23 Bipolar electrolyzer (Adapted from Kroposki et al. 2006)

Monopolar electrolyzers are simpler and their maintenance is easier than that of bipolar electrolyzers, resulting in a lower cost per unit area in the cell. These electrolyzers are generally used for H_2 production of up to 100 Nm^3/h, while bipolar electrolyzers are usually used for over 100 Nm^3/h H_2 production (Carnieletto et al. 2011).

Modern electrolyzers are similar to conventional models, but the electrodes are covered with special coatings, catalysts deposition, and rough surfaces. Some models use membranes separated by Teflon or other materials, and can be operated at temperatures ranging from 80 to 120 °C and yields of 80–90 % (Carnieletto 2011).

2.4 Alkaline Electrolysis (AEL)

From a technological point of view, alkaline electrolyzers are sufficiently well developed and ready to produce renewable hydrogen at significant rates. The equipment is reliable and secure, with total lifetime of up to 30 years, electrode and membrane exchange at every 8 years, operation efficiency ranging between 62 and 82 %, and production capacity from 1 to 760 Nm^3/h (Smolinka et al. 2011). It consists mainly of two electrodes immersed in an aqueous solution of KOH or NaOH (25–30 %). Hydrogen is produced at the cathode and oxygen at the anode. The reactions involved in the process are as follows:

$$\text{Anode:}\ 4OH^- \rightarrow O_2 + 2H_2O + 4e^- \quad (2.29)$$

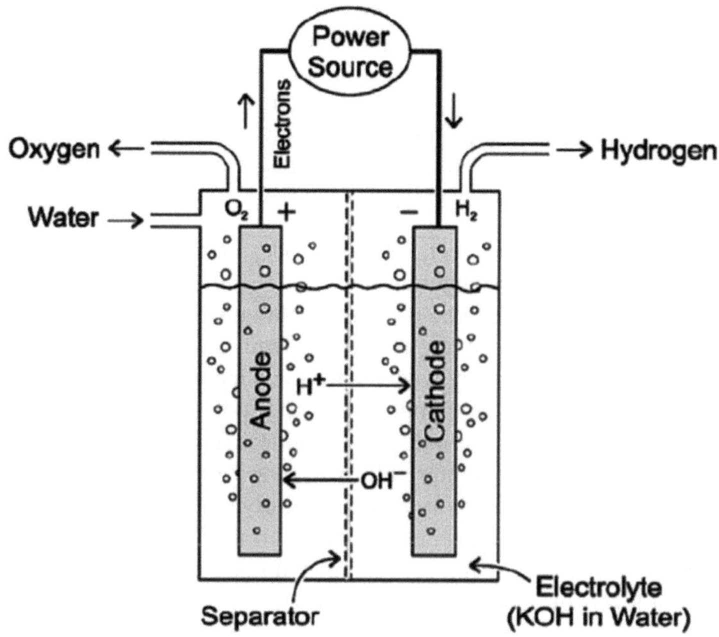

Fig. 2.24 Alkaline electrolyzer scheme (Adapted from Koroneos et al. 2004)

$$\text{Cathode: } 4H^+ + 4e^- \rightarrow 2H_2 \tag{2.30}$$

$$\text{Sum: } 2H_2O \rightarrow O_2 + 2H_2 \tag{2.31}$$

The electrodes are separated by a microporous membrane which is permeable to OH^- ions, but impermeable to gases. The anode is usually made of Nickel or Nickel-coated steel, while the cathode is made of steel coated with different catalysts. The distance between the electrodes is up to 5 mm and the operating temperature is usually limited to 80 °C (Bhandari et al. 2014). Figure 2.24 shows a scheme of the working principle of an alkaline electrolyzer.

There are alkaline electrolyzers operating at low pressures (up to 6 bar) and also at high pressures (from 6 to 30 bar). The advantage of operating at high pressures is that there is no need for a posterior compression of the produced hydrogen, although high pressures reduce the purity of the product due to increased permeability of the membrane in comparison with gases (Bhandari et al. 2014).

The alkaline electrolyzers' energy demand depends on the designing characteristics of the electrodes and their operating conditions. At low pressures, the specific energy demand is between 4.1 and 4.5 kWh/Nm^{3}H$_2$, sometimes reaching 7 kWh/Nm^{3}H$_2$ due to compression. For pressurized electrolyzers, the specific energy demand is between 4.5 and 5 kWh/Nm^{3}H$_2$ (Smolinka et al. 2011).

Improvements in technology have followed two directions. The first one consists in improving the electrolyzer efficiency in order to reduce operating costs associated with electricity consumption. The second one aims to increase the operating current density to reduce investment costs. This situation is most evident in the case of large units in which investment costs are nearly proportional to the surface area of electrolytic cells (Ursúa et al. 2012). Investment costs are widely ranged between $1000 and 2300€/kW (Bertuccioli et al. 2014).

The alkaline electrolyzer has evolved over time. According to Ursúa et al. (2012), a few improvements were:

- Minimizing the space between the electrodes, reducing ohmic losses, and allowing operation with stronger currents.
- Developing new materials to be used as diaphragms instead of asbestos. In this case, the use of ionic-exchange inorganic membranes has been widely developed. Some examples are polyacid antimony membranes impregnated with polymers, porous composite compounds with a polysulfone and ZrO2 (Zirfon) matrix, besides polyphenyl sulfide-based splitters (Ryton).
- Development of high-temperature electrolyzers. Operating temperatures above 150 °C increase the conductivity of the electrolyte and promote the kinetics of the electrochemical reaction under the surface of the electrode. These electrolyzers are intended for producing hydrogen on a large scale, reaching purification levels of up to 99.9 % (Ivy 2004).
- Development of advanced electrocatalyzing materials to reduce the electrode's overvoltage.

Alkaline electrolysis is a mature technology. Several manufacturers like Hydrogenics, Mcphy, Teledyne Energy Systems, among others, have been selling this technology.

2.5 Electrolysis in Acid Medium (PEM Electrolysis)

Known as Proton Exchange Membrane (PEM) or Solid Polymer Electrolyte (SPE) electrolysis, this technology differs from alkaline electrolysis due to the fact that it does not require any electrolytic liquid. In this case, a thin splitting polymer membrane is used, allowing a close proximity of the electrodes. The membrane used in this device is Nafion®, developed by Dupont, which are less than 0.2 mm thick (Ursúa et al. 2012). The electrodes are composed of noble metal alloys, such as Platinum and Iridium. The chemical reactions that occur in a PEM electrolyzer are as follows:

$$\text{Anode: } H_2O \rightarrow \tfrac{1}{2}O_2 + 2H^+ + 2e^- \tag{2.32}$$

$$\text{Cathode: } 2H^+ + 2e^- \rightarrow H_2 \tag{2.33}$$

At the anode, the water is oxidized for producing oxygen, protons, and electrons. The protons pass through the membrane to the cathode where they are transformed

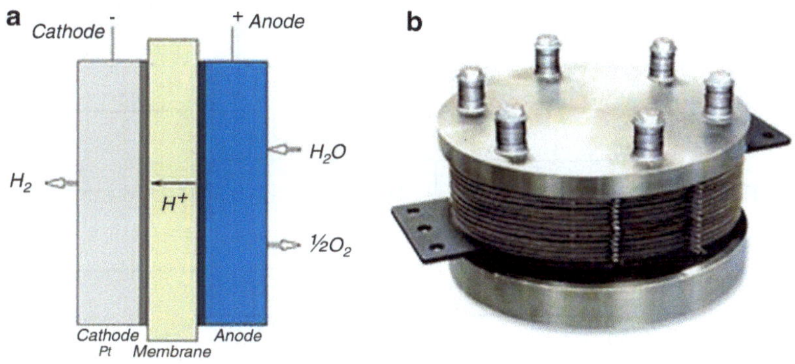

Fig. 2.25 (**a**) PEM electrolyzer operation and (**b**) commercial-type electrolysis battery

into high purity hydrogen, typically over 99.99 % purity. The operating temperature is 80 °C at a pressure of up to 15 bar. The specific energy demand of PEM electrolyzers is between 4.5 and 7.0 kWh/Nm^3H_2, with production capacity ranging between 0.06 and 30 Nm^3H_2/h and operation efficiency ranging from 67 to 82 % (Smolinka et al. 2011; Ursúa et al. 2012).

PEM electrolyzers have emerged to circumvent some difficulties that the alkaline solution presents, such as the significant increase in the corrosion of electrodes (Sorensen 2005). They are not very sensitive to the effects of fluctuations in power supply, which makes them suitable for being applied to energy storage from renewable sources, unlike alkaline electrolyzers which have their efficiency compromised by presenting large inertia in transporting ions (Bhandari et al. 2014). The main problem is the high investment cost associated with the membranes and noble metals of electrodes. It is a technology which is still under development, but there are already several manufacturers, such as ITM Power, ProtonOnSite, and Siemens.

Figure 2.25a illustrates the operation of such electrolyzer and Fig. 2.25b shows a commercial-type electrolysis battery.

2.6 High Temperature Electrolysis (HTEL or SOEL)

SOE electrolyzers (solid oxide electrolyzers) perform electrolysis at high temperatures (600–1000 °C), enabling greater efficiencies than alkaline and PEM models, once the required power supply plummets. However, this process is hindered by the difficulty in finding thermally stable and waterproof materials that can last for a long period of time. The process basically consists in a solid oxide fuel cell working in reverse mode, where water vapor is introduced into the cathode at a high temperature which is reduced to produce hydrogen, and oxide anions are generated and pass through the solid electrolyte to the anode, where they are recombined to

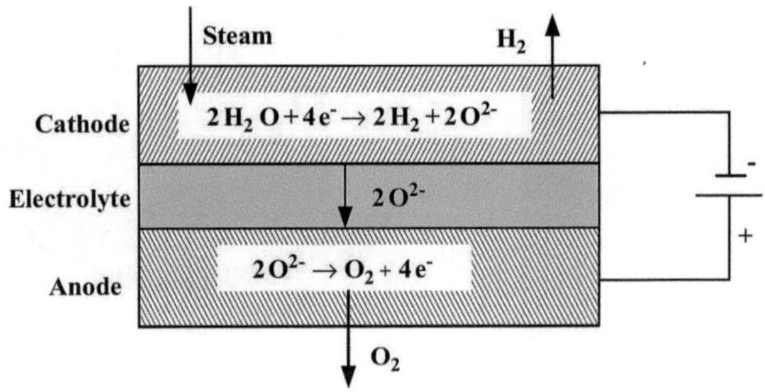

Fig. 2.26 Operation of a solid oxide electrolyzer (Adapted from Wang et al. 2014)

form oxygen. The cathode consists of a hard metal with hard nickel-base particles with yttrium and zirconia (YSZ), where the electrolyte is a solid made of YSZ and the anode is made of perovskite ($CaTiO_3$) (Ursúa et al. 2012). This kind of electrolyzer reduces the required power supply in up to 25 % by using heat from the process that occurs at 1000 °C (Brisse et al. 2008). This feature makes the SOE attractive for producing hydrogen when a high-temperature heat source is available, such as those from nuclear reactors, geothermal energy, and solar thermal energy. Currently, this type of electrolyzer is in its research phase and under development (Bhandari et al. 2014). It is expected that it will be commercially available in 3–5 years at an estimated cost of 2000 €/kW (Bertuccioli et al. 2014). Figure 2.26 shows the diagram of the process.

2.7 Renewable Electricity Sources for Electrolysis

2.7.1 Wind Power

Wind power is mainly produced from solar radiation, since winds are generated by a nonuniform heating of the Earth's surface. According to Dutra (2008), an estimate of the total wind power available around the planet can be made from the hypothesis that approximately 2 % of the solar energy absorbed by the Earth is converted into kinetic energy of winds. This percentage, though it seems small, represents hundreds of times the annual installed power of electricity plants worldwide.

According to the *International Renewable Energy Agency* (Irena et al. 2013), on account of technological advances in wind turbines, the cost of wind power in Latin America has varied from 0.05 to 0.17 US$/kWh over the past few years.

However, the cost of wind power generation is more expensive in Latin America than in other countries that have adopted measures to encourage the construction of wind farms, like Germany, Spain and Denmark. This can be attributed to higher logistic costs for implementing the projects and to the limited number of domestic suppliers of wind turbines, which can be associated with import restrictions on such instruments (Dantas and Leite 2013). The winds that blow on a global and small scale are influenced by different aspects, among which height, roughness, obstacles, and features of the terrain upon which the wind farm was built stand out (Dutra 2008).

According to Furlan (2012), the available wind power is the kinetic energy associated with the air mass traveling at a uniform and constant speed through the wind turbine, which can be calculated according to Eq. (2.34):

$$P_{\text{disp}} = \frac{1}{2}\rho_{\text{ar}} A v^3 \tag{2.34}$$

where:

P_{disp}—power of wind turbine (W)
ρ_{air}—air density (kg/m^3)
A—cross-sectional area of the wind turbine (m^2);
v—wind speed (m/s)

Once the speed at which the air mass is traveling is reduced, the kinetic energy of the wind is converted into mechanical energy through the rotation of the blades. However, the available wind power cannot be fully harnessed by the wind turbine in the conversion of electrical energy. The power that is effectively harnessed by a wind turbine's blades depends on the energy difference between the upstream and downstream of the rotor blades, as shown in Fig. 2.27 (Roberts 2012).

According to Fig. 2.27, generally speaking, it can be considered, under a macroscopic point of view, that wind speed varies from V to V_0 on the plane of the rotor, with an average speed of ½($V + V_0$) at this flow rate. After a few algebraic operations, it can be determined the mechanical power obtained by the rotor, as shown in Eq. (2.35) (Roberts 2012).

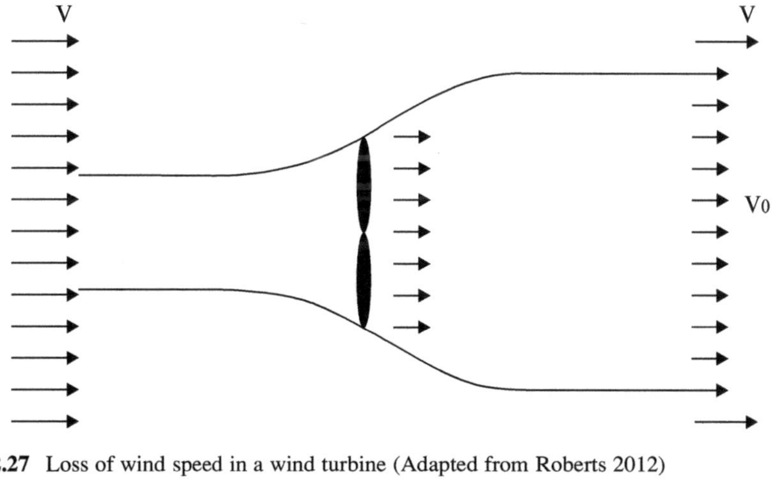

Fig. 2.27 Loss of wind speed in a wind turbine (Adapted from Roberts 2012)

$$P_{mec} = \frac{1}{2}\rho_{ar}Av^3c_p \quad (2.35)$$

where:

P_{mec}—mechanical power of wind turbine (W)

c_p—power coefficient

The concept of power coefficient was introduced by Betz law. The Betz limit indicates that, even when harnessing the most wind power, only a maximum of 59 % can be recovered, which indicates that the maximum c_p (theoretical) is approximately 0.59. According to Roberts (2012), in practice, the maximum efficiency achieved by modern wind turbines ranges from 0.2 to 0.4; the amount of energy that could not be harnessed is dissipated in the downstream wind.

The energy generated in this particular case of electrolysis can be used when there are winds with an average speed of 6 m/s and a minimum speed of 4 m/s (Silveira 2012). Figure 2.28 shows an eolic system that supplies the network with power, in addition to feeding an electrolyzer for hydrogen production.

As illustrated in Fig. 2.28, the hydrogen produced by the electrolyzer is directed to a storage system. The stored hydrogen can be either used in stationary fuel cells to produce electricity for the network, or in vehicular fuel cells of cars or buses.

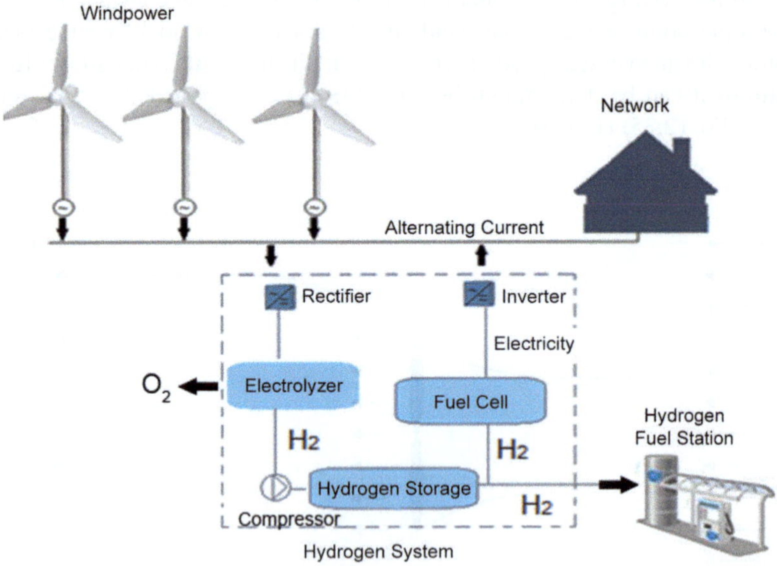

Fig. 2.28 Hydrogen production from wind power

2.7.2 Solar Power

Solar power is the electromagnetic energy from the sun, which is generated via nuclear reactions that, by being propagated through interplanetary space, is emitted onto the Earth's surface. Most other renewable energy sources also depend on the sun as primary source, such as hydroelectric, wind, and tidal power (Roberts 2012). The total solar energy emitted to the Earth's surface yearly is over 10,000 times the annual gross energy consumed by humanity (CRESESB 2012).

According to Silveira (2012), in the Brazilian territory, there is an average solar radiation of 900 W/m^2. A photovoltaic power plant converts 80–180 W/m^2 of this amount into electricity (depending on the type of photovoltaic panel) during 6 h of the day (project parameter, which depends on the position of the sun). You must, however, use banks of batteries to store electricity to be used in periods in which the radiation cannot be harnessed, which pushes up the costs of investment in photovoltaic plants.

The materials used for manufacturing photovoltaic panels are semiconducting elements, where most of them are, on a commercial scale, made of silicon, which are also found in their monocrystalline, multicrystalline or polycrystalline, and amorphous forms (CRESESB 2012). According to Roberts (2012), this preference is due to three factors: silicon is nontoxic and the second most abundant element found in nature, whose technology is fully established.

A single photovoltaic cell generates between 1 and 1.5 W, at a voltage of 0.5–0.6 V, under standard trial conditions (solar radiation of 1 kW/m^2, cell temperature of 25 ° C, and air mass of 1.5 t). In order to obtain voltage and current levels that are suitable to their use, photovoltaic cells are usually arranged serially. Such setup, generally comprising 30 and 36 cells, constitutes a photovoltaic module (Roberts 2012).

The conversion efficiency of photovoltaic cells is measured by the amount of solar radiation that is emitted onto the surface of the cell, which is converted into electrical energy. Table 2.13 shows the efficiencies of photovoltaic panels.

It can be observed in Table 2.13 that the efficiency of photovoltaic panels is 14 % on average, which is relatively low in comparison with the energy generated by wind and hydroelectric turbines.

Solar hydrogen production systems are composed of a set of panels that supply an electrolyzer with direct current electricity. According to Gibson and Kelly (2008), photovoltaic modules are connected to charge controllers and DC/CC converters, which allow the batteries to be fully charged and prevent over-

Table 2.13 Efficiency of silicon-based photovoltaic panels (Reproduced from ANEEL 2011)

Photovoltaic panel type	Efficiency (%)
Monocrystalline silicon	12–14
Concentrated silicon	13–15
Polycrystalline silicon	11–13
Amorphous silicon	3–5

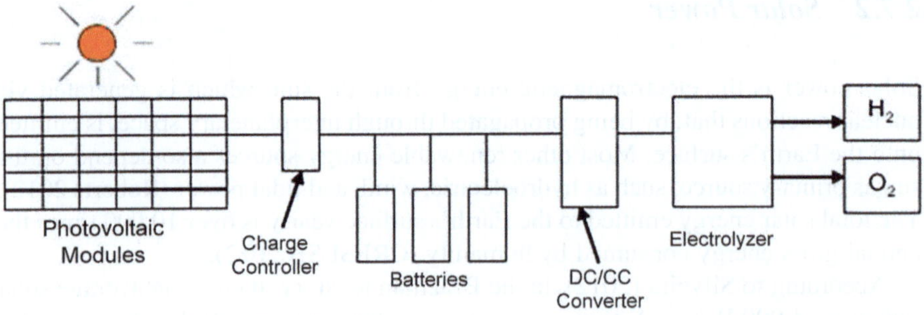

Fig. 2.29 Conventional solar hydrogen production system (Adapted from Gibson and Kelly 2008)

discharging; these converters are needed to supply the typical voltage characteristic of an electrolyzer. A representation of the system can be seen in Fig. 2.29.

When a photovoltaic panel is connected to any equipment, voltage drops below open-circuit due to the internal resistance of the module and equipment. The DC voltage applied to the electrolysis system is limited by the output of the panel circuit, and the voltage and current of the electrolysis process are limited by the operating characteristics of the electrolyzer, hence the need for a DC/DC converter (Gibson and Kelly 2009).

2.7.3 *Hydroelectric Generation from Water Spillage*

Hydropower is based on obtaining electrical energy through harnessing the hydraulic power of a particular segment of a river through the construction of a dam and, consequently, the creation of a reservoir (Pimentel et al. 2012).

For many years, the Brazilian energy generation park has been fundamentally composed of hydroelectric power plants, where the low cost of this type of energy is associated with abundant natural water catchment areas, which enabled the country to reach a prominent position as one of the world's leading energy producers. The electricity derived from hydroelectric power plants accounts for 67.48 % of the installed generation capacity in Brazil, according to the Brazilian Electricity Regulatory Agency (ANEEL 2014). Hydropower generation projects are divided into three groups: the micro hydroelectric power plants (up to 1 MW of installed generation capacity), the small hydroelectric power plants (between 1.1 MW and 30 MW of installed generation capacity), and large hydroelectric power stations (with over 30 MW). According to the Generation Database of ANEEL (2014), Brazil has 1.108 hydroelectric power plants with installed generation capacity of 86.918 MW.

As these power plants are capable of storing only water and not energy, there are periods when the water surplus has to be released by the spillways and a significant amount of water is spilled yearly, thus ceasing energy production. These waters could be harnessed for generating electricity, so as to feed electrolyzers in order to produce hydrogen. Hydrogen can be stored, thus harnessing this energy well can increase the energy efficiency of power plants and minimize water spillage, as well as enabling the introduction of a totally clean energy vector into the energy matrix, thus contributing to green energy generation.

The energy generation potential of Itaipu Dam has been the aim of studies of several Brazilian researchers, with respect to harnessing the power of this amount of water spillage. Power plant managers have been discussing the possibility of building an electrolysis plant since 2003, and have already purchased an electrolyzer, although it has not been installed, yet. Figure 2.30 shows a chart with the difference between the average power production and the amount of energy that could have been harnessed, which has not been generated due to water spillage.

According to Fig. 2.30, it is observed that, in all studied years, a considerable amount of energy ceases to be produced. In 2010, the Dam no longer generated an average power production of 1508 MW, the highest amount since 2006.

Dams suffer from a few drawbacks, which contradict the concept of clean and cheap energy generation. Studies show that Balbina, Tucuruí, and Samuel, the three largest hydroelectric power plants operating in the Amazon region, emit pollutants in the same proportion as coal plants. This may come to you as a surprise, but in the first 10 years of operation of a power plant in the Amazon, organic matter rots on account of being covered by water. This is a very strong process of decay, which acidifies the water and releases methane (CH_4) which is 21 times more harmful than carbon dioxide (CO_2) as regards the greenhouse effect (Azenha 2013).

A hydroelectric power plant is aimed at generating electricity by harnessing the hydraulic power of rivers. In this case, electrolysis would only be justified if the electricity generated from the water that would have been spilled were used, i.e.,

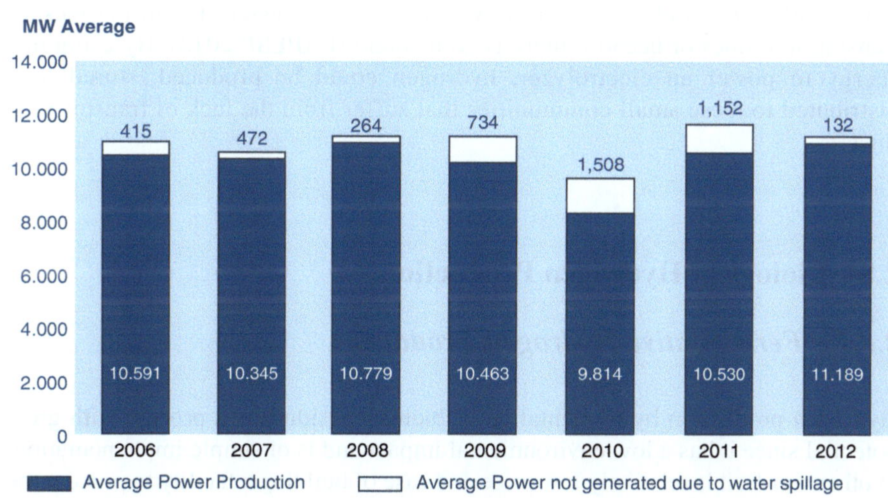

Fig. 2.30 Average power production and water spillage at Itaipu Dam (Reproduced from ITAIPU 2014)

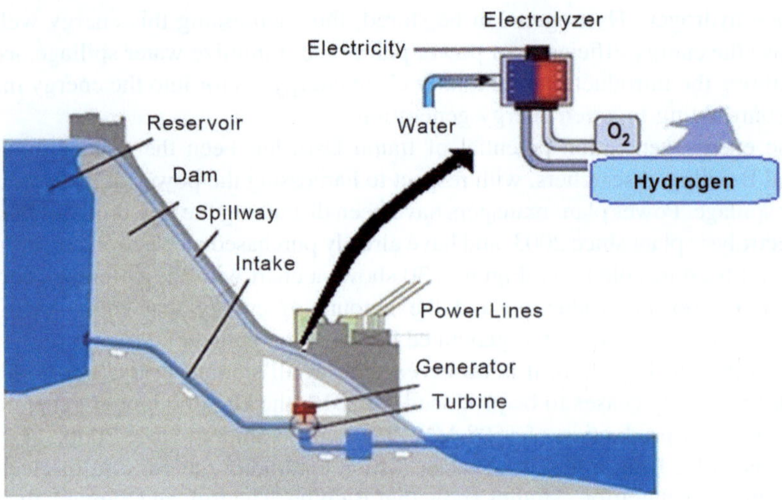

Fig. 2.31 Electrolysis from water spillage in a hydroelectric power plant

using the surplus energy of the plant for hydrogen production. Figure 2.31 shows a system of hydrogen production by electrolysis using hydraulic energy.

According to Fig. 2.31, the water that would pass through the spillway should have been used to power a generator that would, in turn, supply an electrolyzer with electricity, thus producing hydrogen through water electrolysis.

In the case of Brazil, due to the country's large hydroelectric power generation potential, electrolysis by harnessing the water spilled in hydroelectric power plants can be quite promising. As previously noted in Fig. 2.30, Itaipu has enough potential to generate energy from water spillage yearly, which could supply small towns with enough electricity; however, it is not harnessed due to the lack of transmission lines or because there is no demand (FAPESP 2013). By using this energy to power an electrolyzer, hydrogen could be produced, stored, and distributed to these small communities that suffer from the lack of transmission lines.

2.8 Biological Hydrogen Production

2.8.1 Fermentative Hydrogen Production

Hydrogen production by fermentation of biomass residues is a process with great potential since it has a low environmental impact and is of simple implementation, in other words, it is relatively low price and ease of building a facility to process the residues and obtain hydrogen when compared to other processes. It is also important to note a few other advantages of production by this process, such as local integration due to the possibility of adapting to the use of different types of raw material, use of the fermentation residues as fertilizer, and also reducing the economic and environmental impact of fuel transportation (Foglia et al. 2011; Claassen and de Vrije 2006).

The anaerobic fermentation usually takes place in four steps: hydrolysis, acidogenesis, acetogenesis, and methanogenesis.

- Hydrolysis: The first stage of fermentation, where the particulate matter is dissolved, thus being able to cross the cell walls of the bacteria responsible for fermentation. This is due to the excretion of exoenzymes by the bacteria.
- Acidogenesis: The process takes place after the hydrolysis, breaking the molecules into smaller ones inside the cell. As a result of the acidogenesis process, the formation of volatile fatty acids, carbon dioxide production, ammonia, and hydrogen sulfide occurs.
- Acetogenesis: In this stage of the process, the previous stage products are oxidized. They are digested, forming hydrogen, acetic acid, and carbon dioxide.
- Methanogenesis: This is the final stage of the anaerobic digestion process, with the formation of methane, water, and carbon dioxide.

The fermentation process is shown in Fig. 2.32 below:

Some factors interfere with the fermentation process. Among which are the influence of temperature, raw material composition, particle size, pH, among others.

There are several materials that can be used as feedstock for hydrogen production by fermentation. The most commonly used ones are rich in lipids, proteins, and carbohydrates, the latter being the preferred source for the fermentation process.

2.8.2 Enzymatic Hydrogen Generation

According to Woodward et al. (2000a, b), marine thermophilic anaerobes are able to produce hydrogen and carbon dioxide. For such a purpose, the hydrogenase from *Pyrococcus furiosus* should be coupled in vitro with glucose dehydrogenase in order to produce one mole of hydrogen per mole of glucose (Woodward et al. 1996; Inoue et al. 1999). Since the same NADP+ cofactor, i.e., nicotinamide adenine dinucleotide phosphate (Ma et al. 1993, 1994), was selected to be used for both enzymes, it is recycled while the glucose substrate is being oxidized for molecular hydrogen production. This system can also be used to produce hydrogen from sucrose or cellulose. Thus, appropriate enzymes such as invertase or cellulase must be added (Woodward and Orr 1998; Woodward et al. 2000a, b).

The anabolic pentose phosphate pathway (PPP) is commonly found in most organisms. Its oxidative branch comprises two enzymes: glucose 6-phosphate dehydrogenase and 6-phosphogluconate dehydrogenase. Such enzymes together produce 2 mol of NADPH and 2 mol of ribose 5-phosphate from one mole of glucose-6-phosphate. The non-oxidative branch makes one and two carbon

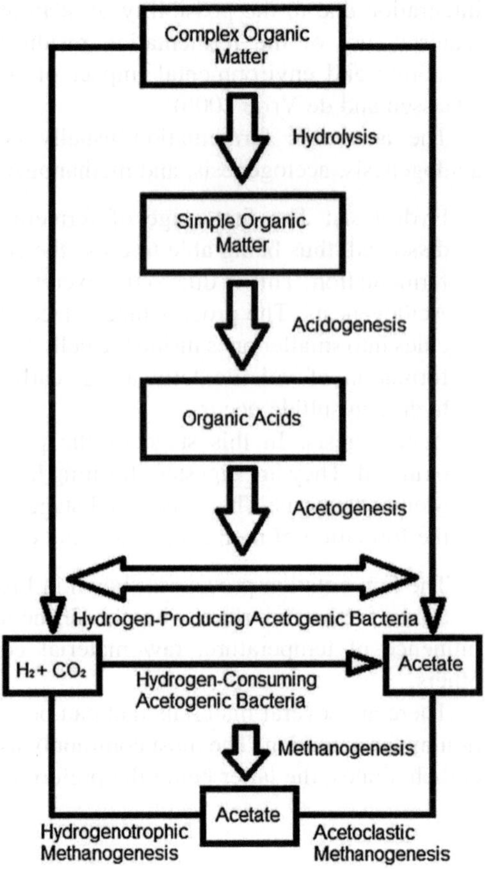

Fig. 2.32 The anaerobic digestion process (Adapted from Santos 2004)

transfers, which convert pentoses into fructose 6-phosphate and glyceraldehyde 3-phosphate. Given that fructose 6-phosphate can be isomerized into glucose 6-phosphate through phosphohexose isomerase (glucose 6-phosphate isomerase), they can be recycled back into NADPH for further hydrogen production. The addition of hydrogenase theoretically obtains 12 H_2 mol per 1 mol glucose 6-phosphate. A yield close to the aforementioned theoretical value (11.6 mol H_2 per 1 mol glucose 6-phosphate) was reached by combining PPP mesophilic PPP enzymes and *P. furiosus* hydrogenase, which is depicted in Fig. 2.33 (Woodward et al. 2000a, b).

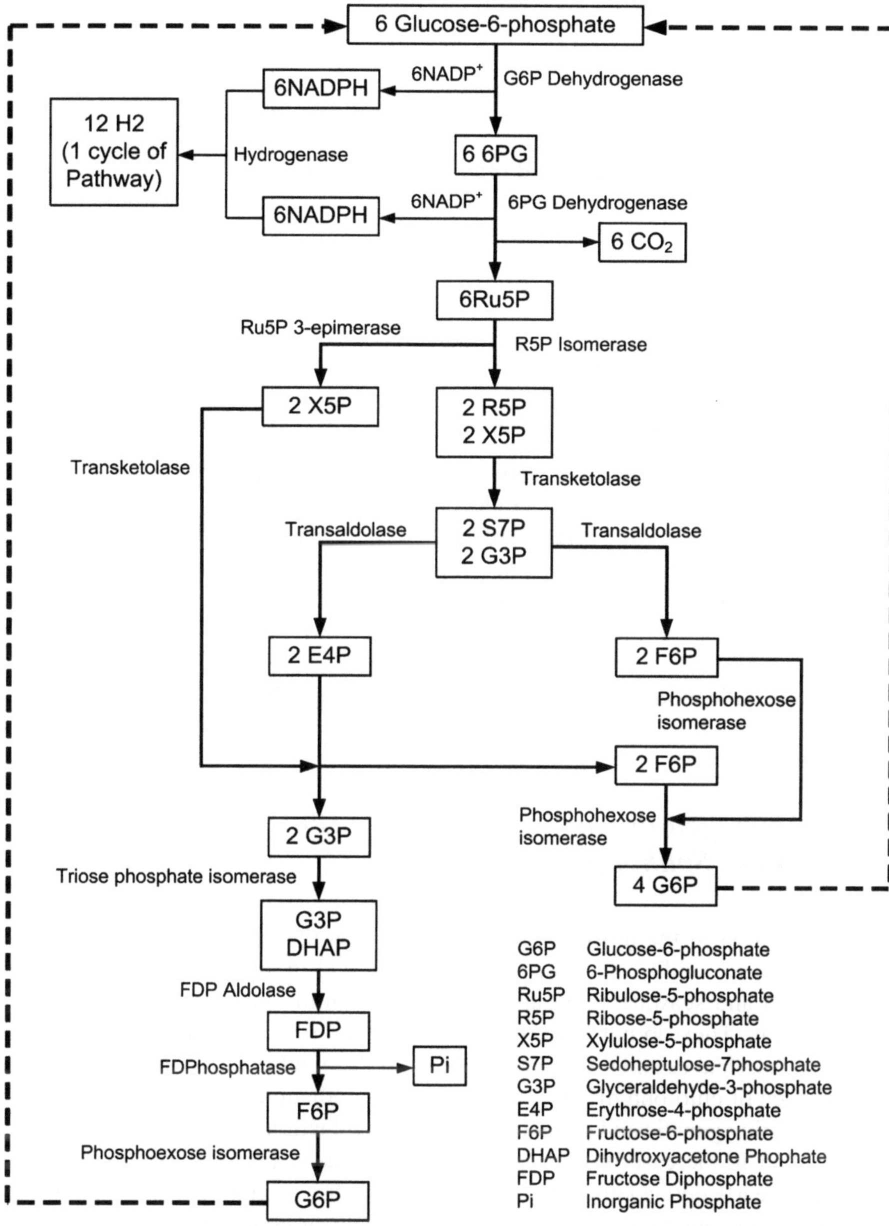

Fig. 2.33 Hydrogenase and enzymes of the PPP for hydrogen production (Adapted from Woodward et al. 2000a, b)

2.8.3 *Biocatalyzed Electrolysis*

According to Rozendal (2007), biocatalyzed electrolysis is a feasible hydrogen production method, since it is:

- Capable of converting dissolved organic compounds
- Able to deal with the endothermic nature of the conversion reactions of many dissolved organic compounds into hydrogen in a more practical way than using sunlight
- Robust and flexible, as it will be designed for hydrogen production from wastewater
- Based on mesophilic mixed cultures that grow along the process through natural selection
- Able to convert a wide range of organic substrates

A biocatalyzed electrolysis system makes use of electrochemically active microorganisms which are able to transfer electrons from the inner to the outer part of the cell and capable of growing on the surface of the electrode during its use as electron receiver for the oxidation of organic compounds in which it is immersed. Since electrochemically microorganisms release electrons at a high energy level while immersed, a low potential will appear on the electrode, thus producing hydrogen (Rozendal 2007).

The schematic model of a fuel cell based on this principle is shown in Fig. 2.34.

2.8.4 *Hydrogen Production from Algae (Photobiological Water Splitting)*

Algae are amongst the most resistant organisms on the planet, which are able to grow in diverse conditions. They are typically found in damp places or bodies of water (fresh and seawater) and are, therefore, common in terrestrial and aquatic environments. They do not have the various structures that characterize terrestrial plants, such as leaves and roots, and are different from other microorganisms due to the presence of chlorophyll. They possess photosynthetic capacity in a single cell, thus facilitating genetic and metabolic research conduction in a much shorter period of time than in the case of conventional plants. The major components of green algae are: clearly defined nucleus, cell wall, chloroplasts containing chlorophyll and other pigments, pyrenoid, stigma, and flagella (Pelczar et al. 2008).

Biological hydrogen production from algae has been considered as being sustainable and advantageous in comparison with thermochemical and/or electrochemical processes, due to the purity of hydrogen (over 98 %), use of a simple solar reactor, and the consumption of CO_2 during their growth phase; moreover, they are capable of fixing CO_2 more efficiently than terrestrial plants (in about ten times) as regards the amount of biomass per m^2 (Usui and Ikenouchi 1997).

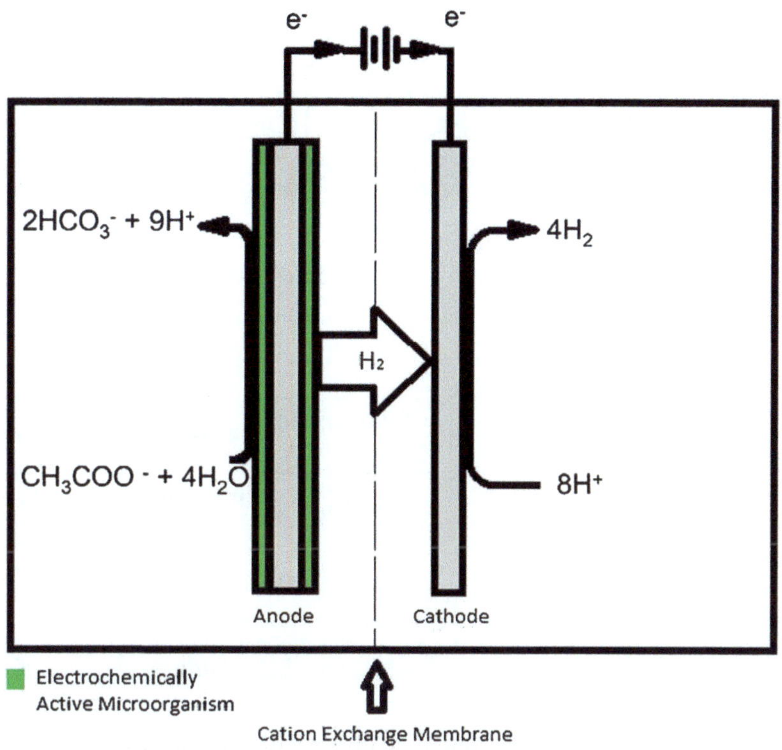

Fig. 2.34 Microbial fuel cell (Adapted from Logan et al. 2006)

In comparison with different types of algae for hydrogen production, the microalgal strain *Chlamydomonas reinhardtii* (*C. reinhardtii*) presents one of the simplest lifecycles, of easy manipulation with rapid growth, and one of the best hydrogen production rates (Hemschemeier 2005; Ministry of Agriculture and Agroenergy 2006). Hydrogen production from this strain is usually performed by direct biophotolysis, as shown in Fig. 2.35 (Ahmed et al. 2011; Hallenbeck and Benemann 2010; Das and Veziroglu 2008).

Direct biophotolysis occurs in the presence of light and under anaerobic conditions. According to Fig 2.35, in photosystem 2 (PS II), light energy is captured by pigments. These pigments are usually chlorophyll and carotenoids attached to proteins to form a complex which absorbs photons. This absorbed energy is transferred to the reactive core of this photosystem, and the energized electrons are transferred by the ferredoxin to the hydrogenase. To replace these lost electrons, the reactive core withdraws electrons from two water molecules, resulting in the detachment of O_2 and release of four hydrogen ions ($4H^+$) (Lehninger et al. 2010).

Hydrogenase is activated only under anaerobic conditions, and it uses the electrons received by ferredoxin to reduce protons (H^+) to hydrogen molecules (H_2) (Lehninger et al. 2010).

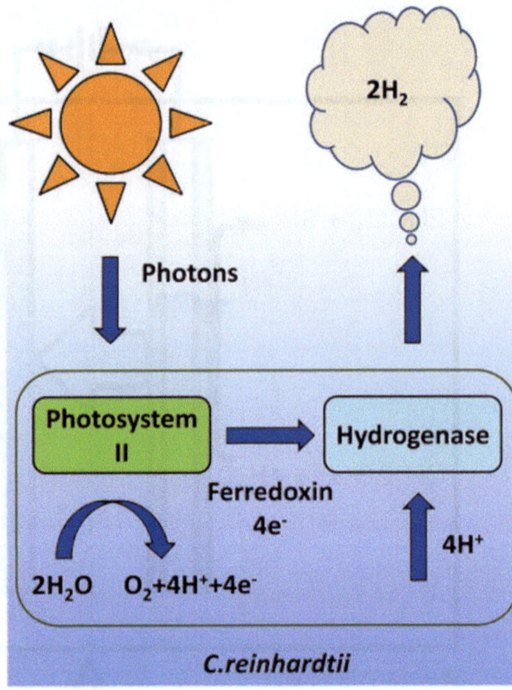

Fig. 2.35 Hydrogen production by direct biophotolysis from *C. reinhardtii* (Adapted from Tamburic et al. 2011)

After years of research, it was observed that in closed cultures, the deprivation of sulfur in these organism strains results in oxygen consumption and anaerobiosis in the culture medium. Thus, the medium microorganisms are performing an alternative photosynthesis, which uses light and hydrogenase for H_2 production (Tao et al. 2008).

2.8.4.1 Hydrogen Production Process from the Strain *Chlamydomonas reinhardtii*

The experimental photobiological hydrogen production from green microalgae *C. reinhardtii* occurs in two steps. The first one refers to culture the algae and the second one to hydrogen production. Thus, it occurs in two different reactors, as shown in Fig. 2.36.

In the first reactor, the algae grow in the presence of light, with normal concentration of sulfur, where pH and temperature should remain close to 7 and 25 °C, respectively (Effendi et al. 2005; Haag et al. 2007). The specific composition of the medium for the growth of this microalgal strain is presented in Table 2.14.

At room temperature and pressure of 1 atm (Levin et al. 2004), the content of the first reactor is transferred to the second reactor after being centrifuged or diluted (10 % v/v), which contains a medium without sulfur and whose composition is

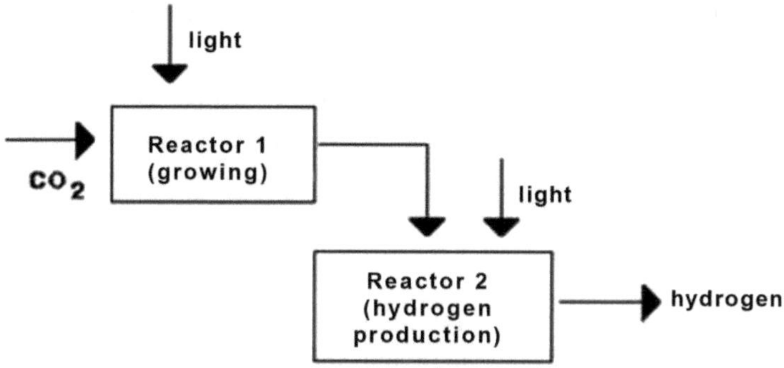

Fig. 2.36 Hydrogen production from green microalgae

Table 2.14 Composition of the *C. reinhardtii* growth medium: Tris-acetate-phosphate (TAP) (Adapted from Xu et al. 2004)

Medium prior to TAP		TAP	
15 g	NH_4Cl	25 mL	Medium prior to TAP
4 g	$MgSO_4{\cdot}H_2O$	0.375 mL	Phosphate solution
2 g	$CaCl_2{\cdot}H_2O$	1 mL	Trace elements (EDTA, Zn, B, Mn, Co, Cu, Mo and Fe salts)
Dissolve in 1 L of water		1 mL	Acetic acid
		2.42 g	Tris-aminomethane
		Dissolve in 1 L of water	

Table 2.15 Hydrogen production medium composition without sulfur (TAP-S) (Adapted from Xu et al. 2004)

Medium prior to TAP-S		TAP-S	
15 g	NH_4Cl	25 mL	Medium prior to TAP-S
4 g	$MgSO_4{\cdot}H_2O$	0.375 mL	Phosphate solution
2 g	$CaCl_2{\cdot}H_2O$	1 mL	Trace elements (sulfur salts replaced by chloride salts)
Dissolve in 1 L of water		1 mL	Acetic acid
		2.42 g	Tris-aminomethane
		Dissolve in 1 L of water	

presented in Table 2.15. The media of the first and second reactor are known as TAP and TAP-S, respectively.

In Tables 2.14 and 2.15, it is observed practically the same composition for both media, and the only difference is in the absence of sulfur in the TAP-S medium. Thus, sulfur salts are replaced by chloride salts. For the purpose of media formation, a TAP or TAP-S solution is firstly made. After this step, 25 mL of this solution, 0.375 mL of the phosphate solution, 1 mL of trace elements, 1 mL of acetic acid, and 2.43 g of Tris-aminomethane are dissolved in 1 L of water, thus obtaining the media for algae growth (TAP) and hydrogen production (TAP-S).

2.8.5 Photobioreactors for Hydrogen Production from Algae

Microalgae can be cultured through open systems (ponds) or in closed systems, called photobioreactors (PBR).

2.8.5.1 Open Systems

The greatest advantage of open systems is its simplicity, which results in low production and operation costs (Singh and Sharma 2012). The tanks are usually called raceway ponds (Fig. 2.37) which are filled with natural waters, made of concrete or simply dug into the ground and coated with plastic to prevent the soil from absorbing the liquid, and their depth is limited so that the algae can be exposed to sunlight. Although this method is the simplest of all cultivation techniques, it presents some drawbacks: out-of-control condition, being highly vulnerable to contamination by other microorganisms and lower biomass productivity.

Generally speaking, the productivity of these systems is 0.5 g L-1d-1, well below the levels reached by photobioreactors because high cellular density cannot be attained. An average depth of 20 cm results in insufficient agitation, as well as leading to low rates of gas-liquid mass transfer and an increase in residence time of the culture at the bottom of the pond (dark zone). In order to overcome the problems associated with an open system, researchers usually select closed systems of monocultures (Singh and Sharma 2012).

Fig. 2.37 Open system for algae cultivation (raceway ponds) (Reproduced from the Sahara Forest Project 2013)

2.8.5.2 Photobioreactors

Photobioreactors are closed culture systems, which are mainly applied in the controlled production of biomass and hydrogen production processes. Despite their elevated costs and design limitations, their mass transfer rate, due to the tendency to accumulate O_2, offer several advantages over open systems (Singh and Sharma 2012), which are:

- Minimizing contamination
- Offering better control of conditions, such as pH, temperature, light, and CO_2 concentration
- Avoiding water evaporation
- Favoring cell concentration

There are several types of photobioreactors, but the most commonly used ones are the tubular and plate photobioreactors. These are systems aerated by gas injection and usually have an agitation mechanism. They can be aligned horizontally, vertically, or inclined, as shown in Fig. 2.38.

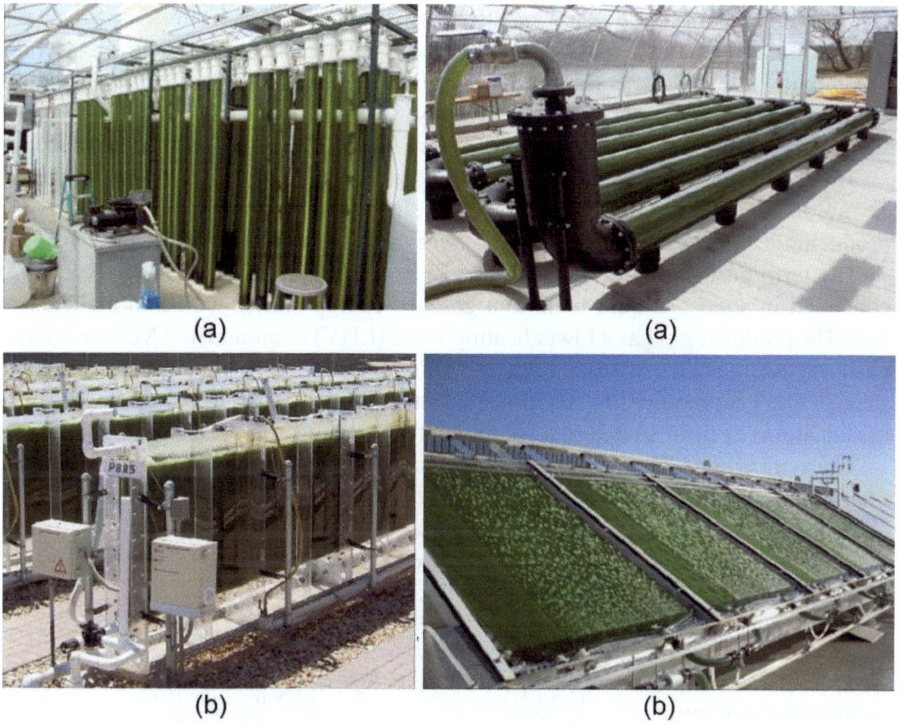

Fig. 2.38 Tubular (**a**) and plate photobioreactors (**b**) (Reproduced from Kwietniewska et al. 2012; University of Nevada 2013; Arizona State University 2013; Energy Options 2013)

According to Burgess et al. (2011), plate photobioreactors are between 1 and 5 cm thick, and their height and length may vary. They have high surface-area-to-volume ratio and, therefore, achieve high energy efficiency when compared to tubular photobioreactors (Cortés 2009). Tubular photobioreactors have a maximum diameter of 10 cm, and also present a large illuminated area. Their main disadvantage is the gradient difference in pH, O_2, and CO_2 dissolved along the pipeline.

All closed systems present advantages and disadvantages, therefore, choosing the type of photobioreactor depends on the process, product to be obtained, species involved, economic assessment, and installation region, as well (Sugai 2012).

2.9 Other Hydrogen Production Processes

2.9.1 Gasification

2.9.1.1 Gasification Theory

Biomass gasification technologies have been used since 1800s (Reed and Das 1998), and the first application occurred in gas production from coal. Since the 20s, gasification has been used to produce synthetic chemicals. The most commonly known technique is the Fischer-Tropsch process for the production of liquid hydrocarbons from synthesis gas. Currently, some gasifiers using coal as fuel are still in operation, but the number of gasifiers fueled by biomass is still limited.

The gasification technology for obtaining liquid and gaseous fuels from biomass as raw material has gained new impetus, since it represents a breakthrough for producing a sustainable alternative of energy carriers (Van der Meijden 2010).

The term gasification is applied to processes that aim at transferring the energy contained in solid or liquid fuels to a gaseous energy carrier with minimal energy loss. The producer gas has a lower heating value (LHV) of around 4–7 MJ/Nm^3 (if the gasifying agent is air), or from 10 to 18 MJ/Nm^3 (if the gasifying agent is oxygen or steam). The heat required for the process can be supplied by burning part of the fuel (gasification) or from an external source (indirect gasification) (Sánchez et al. 2008).

The syngas can have different applications, depending on its quality: for starting internal combustion engines, gas turbines, direct burning, and synthesizing chemical components (Andrade 2007). General specifications that the gas must meet so that it can be applied in modern engines were standardized in the late 1980s, and are reported in Table 2.16 (Van de Beld 2001).

Table 2.16 Gas specifications for internal combustion engine use (Adapted from Van de Beld 2001)

Parameters	Unit	Acceptable	Great
Gas LHV	kJ/Nm^3	>2500	>4200
Particulate matter content	mg/Nm^3	<50	<5
Particle size	μm	<10	<1.0
Tar content	mg/Nm^3	<500	<100
Acetic acid	ppm	<50	0
Hydrochloric acid	ppm	<500	0

2.9.1.2 Gasification Zones and Main Reactions

Gasification is the conversion of solid biomass into a gas fuel (and a portion of sensitive heat), whereas in combustion, the heat capacity of the solid is completely transformed into sensitive heat. Gasification provides some advantages when compared to direct combustion, such as the resulting gas presents better combustion properties than solids, its combustion process is simpler to manage and requires a smaller amount of air with no particulate emissions, thus resulting in less air pollution and better management of the plant.

The chemical composition of bagasse, once it is dry, can be regarded as being approximately constant and equal to $CH_{1,488}O_{0,7}$.

The overall reaction of the gasification of bagasse with air can be schematically described by Eq. (2.36) (Pérez et al. 2014):

$$\begin{aligned} & CH_{1,488}O_{0,7} + wH_2O + mO_2 + 3.76mN_2 \\ & \quad \Leftrightarrow 0.538H_2 + 0.669CO + 0.311CO_2 + 0.32H_2O + 0.02CH_4 \\ & \qquad + 3.76mN_2 \end{aligned} \tag{2.36}$$

After the biomass dehydration and pyrolysis process, the products react with oxygen in a temperature ranging from 1000 to 1200 K (Reed and Das 1998). Initially, the combustion reaction occurs as follows:

$$C + O_2 \Leftrightarrow CO_2 \quad \Delta H^0 = +406.0\,kJ/mol \tag{2.37}$$

Afterwards, a chain of reactions occur, where the main ones are the Boudouard reaction of carbon conversion with CO_2 and steam.

$$C + CO_2 \leftrightarrow 2CO \quad \Delta H^0 = +172.6\,kJ/mol \tag{2.38}$$

$$C + H_2O \Leftrightarrow CO + H_2 \quad \Delta H^0 = +131.4\,kJ/mol \tag{2.39}$$

These reactions occur slowly, thus controlling the speed of the gasification process. The successive reactions are:

$$CO + H_2O \Leftrightarrow CO_2 + H_2 \quad \Delta H^0 = -41.2\,kJ/mol \tag{2.40}$$

$$C + 2H_2 \Leftrightarrow CH_4 \quad \Delta H^0 = -75.0\,kJ/mol \tag{2.41}$$

In the first one, hydrogen is produced by the water-gas shift reaction, and methane production occurs on the next one. In the gasification zone, the following secondary reactions occur:

$$C + \tfrac{1}{2}O_2 \Leftrightarrow CO \tag{2.42}$$

$$CO + \tfrac{1}{2}O_2 \Leftrightarrow CO_2 \tag{2.43}$$

$$H_2 + ½O_2 \Leftrightarrow H_2O \tag{2.44}$$

$$CH_4 + O_2 \Leftrightarrow CO_2 + 2H_2O \tag{2.45}$$

The kinetics of these reactions is difficult to be determined in the actual conditions of gasification because the process is usually studied as a whole.

It is clarified that this theory is not always valid for all types of gasifiers, thence it is necessary to consider that the residence time of all species should be sufficient to reach a state of thermodynamic equilibrium in the system. Ideally, there should be a complete conversion of biomass, but, in fact, it does not happen.

The obtained product typically contains a variable amount of tar and hydrocarbons. The gasification occurs at high temperatures, where there are only a few stable combinations of produced elements, which are: CO, CO_2, CH_4, H_2, H_2O, and a few hydrocarbons (C_xH_y).

The gas produced in the gasification process also contains tar. Tars are heavy hydrocarbons, which can cause incrustation problems when the gas is cooled; it can also contain other pollutants, such as H_2S, thiophene, NH_3, HCl, HCN, COS (Carbonyl sulfide), and solid particles, that have to be removed before the gas is used in some applications (Van der Meijden 2010).

A widely used approach consists in expressing the concentration of different species as a function of temperature and consumed oxygen; the latter is represented by the equivalence ratio (ER) (Reed and Das 1998):

$$\mathrm{ER} = \frac{\text{oxidizer mass/dry fuel mass}}{\text{oxidizer mass /fuel mass*}} \tag{2.46}$$

*mass stoichiometric ratio for complete combustion.

This parameter, arbitrarily introduced along with temperature, efficiently describes the gasification process. When $0 \leq \mathrm{ER} \leq 0.2$, pyrolysis occurs with an oxygen deficit; however, if $\mathrm{ER} \geq 1$, a complete combustion occurs, with a theoretical or higher amount of air. As for the gasification process, the ER generally varies between 0.2 and 0.4 (Reed and Das 1998). Table 2.17 shows the stoichiometric air calculation for the combustion of 1 kg of bagasse with the composition considered in this study.

2.9.1.3 Types of Gasifiers and Main Operation Feature

The characterization of gasifiers can be made according to different properties, such as bed temperature and type, relation between the gasifying agent and fuel, pressure and ash formation. As regards temperature conditions, reactors can be classified into high-temperature gasifiers (typically from 1500 to 1800 K) which produce a gas with higher heating values (10–18 MJ/Nm^3) (*syngas*) and average-temperature gasifiers (typically 1123 K) which produce a fuel gas with lower heating values (4–7 MJ/Nm^3). The *syngas* contains low hydrocarbon content as methane, but the

Table 2.17 Stoichiometric air determination according to the aforementioned bagasse composition

Component	[kg/100 kg of bagasse]	Molar mass component [kg/kmol]	Component substance quantity [kmol]	Product	[kmol O / kmol component]
C	36.56	12	3.05	CO_2	6.09
H	4.37	1	4.37	H_2O	2.19
S	0	32	0.00	SO_2	0.00
O	31.25	16	1.95	–	–1.95
N	0.16	14	0.01	N_2	0.00
W	25	18	1.39		0.00
O stoichiometric moles (kmol)/100 kg solid					6.33
O_2 stoichiometric moles (kmol)/100 kg solid					3.16
Oxygen volume per kg of solid Nm^3/kg solid					0.71
Oxygen mass per kg of solid (kgO_2/kg solid)					1.01
Considering air composition with 21 % oxygen					
Required air (kmol)/100 kg solid					15.07
Air volume per kg of solid (Nm^3Ar/kg solid)					3.37
Air mass per kg of solid (kg Ar/kg solid)					4.36

gas from average-temperature gasifiers contains high hydrocarbon content (primarily CH_4, C_2H_4, and C_6H_6), depending on the type of reactor, biomass, and gasification agent being used (Sánchez et al. 2008).

The most efficient type for syngas production is the entrained-flow gasifier. Such gasifiers, which operate on coal and technologies developed by Shell, General Electric and Siemens, are commercially available in the order of tens and hundreds of thermal megawatts. They are mainly operated at high pressure conditions (typically from 3000 kPa) to maximize the efficiency of the process in the reactor. Typically, O_2 (diluted with steam) is used as gasification agent. The gasifier always works above the ash fusion temperature in order to keep it in its liquid form. The syngas produced by the gasifier is subsequently cooled so as to solidify the gas along its purification process (Vreugdenhil et al. 2009).

Average-temperature gasifiers can be classified based on their bed, such as fixed-bed and fluidized bed gasifiers. Fixed bed gasifiers can be classified into co-current gasifiers (*Downdraft*) and counter-current gasifiers (*Updraft*). Both are commonly used for biomass gasification. Figure 2.39 shows the basic principles of a typical operation of updraft (a) and (b) downdraft gasifiers.

Downward gasifiers are widely used for small-scale cogeneration. Their typical input thermal powers are between 100 and 1000 kW. The fuel is biomass with moisture content of around 20 %. The advantage of this technology is that the produced gas contains low tar and particulate matter content, and its technology is relatively simple.

A fixed-bed gasifier also requires a uniform fuel distribution for a continuous and reliable operation. The carbon content of ashes at the bottom of gasifiers is usually high (>10 wt%) because the fuel conversion is not complete along the gasification process. The downdraft technology is not suitable for being applied on an industrial

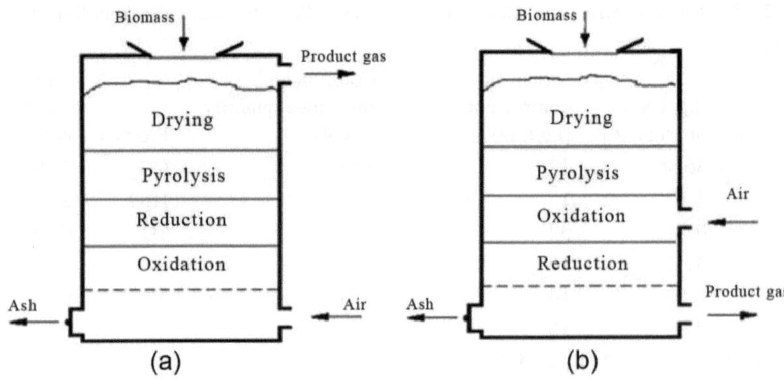

Fig. 2.39 Fixed-bed gasifiers operation: (**a**) updraft and (**b**) downdraft

scale because, among other reasons, it requires an increase in bed diameter; such an increase raises the risk of channeling, thus producing a homogeneous distribution of oxygen in the gasification section and decreasing the efficiency of the process.

Updraft gasifiers are more suitable for being applied on an industrial scale and are less sensitive to variation in moisture content and biomass shape, but produce a large amount of tar in their traditional design. If tar were completely removed, the gas could be combusted in a gasoline engine. Tar is normally removed once it is combined with water, and this water flow requires a thorough purification process before being released into a sewer system. The overall efficiency of the updraft reactor can be increased due to a high rate of carbon and gas conversion, moreover, because the syngas is released at a low temperature from the gasifier. The tar removal and water cleaning make the process complex and too expensive for being applied on a small scale (<1 thermal MW). An example of a successful updraft reactor can be found in a plant called Harbøøre in Denmark (Paisley et al. 1991; Van der Meijden 2010).

Fixed-bed biomass gasifiers (updraft and downdraft) are operated in dry mode, which means that ashes do not reach their melting point. This result is achieved by keeping the operating temperature of the reactors below the melting point of ashes. Both types of gasifiers generally use air as gasifying agent. Fixed-bed gasifiers are not seen as an option for producing synthesis gas from biomass because of their inability to produce a nitrogen-free gas when using air as gasifying agent, in addition to the amount of tar in the fuel gas.

2.9.1.4 Gasification in Fluidized Bed Theory

Fluidized bed gasifiers can be operated in such a way as to produce a gas which is either nitrogen-free or low in nitrogen. The technology is suitable to be expanded to several tens of MW. These gasifiers have been widely used for several years, and their advantages over fixed-bed gasifiers lie in the uniform temperature that is

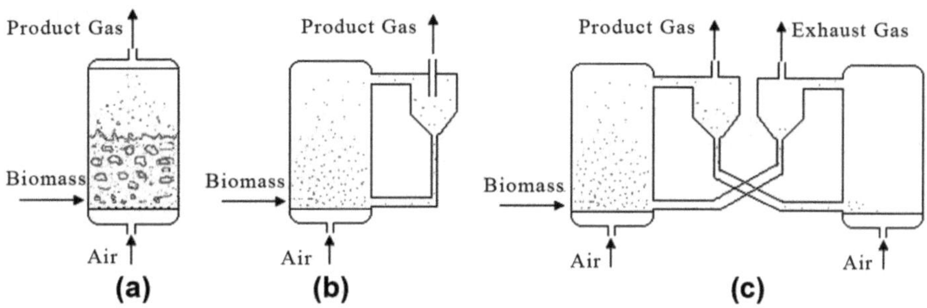

Fig. 2.40 Fluidized bed gasifiers: (**a**) bubbling fluidized bed, (**b**) circulating fluidized bed and (**c**) indirect gasification

reached in the gasification section and the greater speed of the gasification reaction occurrence (Raj and Goic 2011).

Fluidized bed reactors are those in which the gasifying agent circulates through its interior at such a rate that the bed is found in a fluidization state, where there are various conditions that intensify the energy and material transfer between the fuel and the gas. In this type of system, the gasifying agent enters through the bed of particles at a rate which is enough to keep them suspended. The bed is heated and, when high temperatures are reached, the fuel particles are introduced, reaching bed temperature quickly. Biomass gasification in fluidized-bed reactors requires, in general, small fuel particles (<6 mm) and a moisture content which is preferably less than 30 % of its weight (Basu 2006).

Fluidized bed gasifiers can be divided into three main categories: bubbling fluidized bed (BFB), circulating fluidized bed (CFB), and indirect gasification with dual fluidized bed. All fluidized bed gasifiers use an inert material in their bed, which can be ordinary sand, fuel ashes, or a catalytically active bed material, such as dolomite or olivine. The purpose of the bed material is to conduct and distribute heat through the gasifier, so as to homogenize the temperature, mix the fuel with the gasifying agent and the syngas, and, in the case of a catalytically active material, reduce the concentration of tar. Figure 2.40 shows the basic principles of the three types of fluidized bed gasifiers (CIRCE 2005; Van der Meijden 2010).

In a bubbling fluidized bed gasifier (BFB), the fuel is usually fed at the base of the bed. The bed material is fluidized by a gasifying agent (air, oxygen, or a mixture of steam and oxygen) that enters the gasifier by means of injectors distributed at the bottom of the reactor. In such reactors, the fluidization speed of the gasifying agent is low because there is a significant movement of the solid; the gasifying agent's typical speed is 1 m/s. Oxygen is used in the bed for combusting part of the gas and/or coal to produce the necessary heat for triggering the typical thermochemical reactions of the process. These gasifiers are usually applied on a scale of less than 10 MW (thermal gasifiers). The reason for this capacity limitation is that there must be a good distribution of the fuel along the bed, which becomes more difficult as the reactor diameter increases. Table 2.18 presents a comparison between the circulating and bubbling fluidized bed gasifiers (Williams 1996).

Table 2.18 Comparison between circulating and bubbling fluidized bed reactors (Adapted from Williams 1996)

Fluidized bed reactor	Temperature [K]		Biomass	Feed	Gasifying agent	Tar content
	Reaction	Outlet				
Bubbling (BFB)	973–1273	973–1073	Wood chips, corn cobs, rice husk	Directly into the bed area	Lower part of the gasifier	Relatively high
Circulating (CFB)	973–1273	873–1073	Bagasse, wood chips, rice husk, sawdust	Directly into the bed area	Lower part of the gasifier	Low

Circulating fluidized bed gasifiers (CFB) have higher fluidization speeds and, for this reason, the bed material is dragged, thus requiring circulation to maximize fuel conversion. The speed of a typical gasifying agent is usually between 3 and 10 m/s, which is much higher than in the case of BFB. The dragged bed material and the fuel particles that have not been converted completely are removed from the gas through a cyclone separator or another device. The particles will typically return to the bottom of the gasifier through a nonmechanical valve. Such nonmechanical valve can be a tube that also has the function of preventing gas leaks at the bottom of the solid outlet in the cyclone separator. The company Foster Wheeler successfully implemented this type of gasifier on a commercial scale in a project developed in Lahti, in Finland, and in Ruien, Belgium (Paisley et al. 1991; Van der Meijden 2010).

On indirect gasification, the biomass gasification separation and the remaining coal is combusted in separate reactors, as shown in Fig 2.40c. The biomass fed into the first reactor is converted into gas and coal. The heat required for the process is conducted through the circulating bed material, and it is originated from the second combustion reactor. The coal and the bed material are separated from the gas by a separating device (for example, a cyclone separator); the syngas leaves the gasifier and goes through a purification process.

2.9.1.5 Advantages of Fluidized Bed Gasification

The main advantages of fluidized bed reactors include: better control of temperature and reaction rates, high specific capacity, potential for dimensioning to magnitudes on the order of hundreds of megawatts, easy adaptation to changes in biomass, and low efficiency losses caused by particles that may not react. These types of reactors also present low sensitivity to variations in fuel moisture (Zainal 2010).

2.9.1.6 Disadvantages of Fluidized Bed Gasification

The main disadvantages of fluidized bed gasifiers are particle losses which lead to a decrease in the process efficiency. In addition, the internal walls of the reactor and the elements involved in it cause wear from the high relative speed of the fluid and friction with the sand and particles. Furthermore, these reactors show moderate and high levels of tar and particles in the exhaust gases and, therefore, require a complex system of gas cleaning and residual water treatment. Another drawback is that the carbon conversion rate is lower than that of fixed bed gasifiers (Zainal 2010).

2.9.1.7 Gasification Theory on Entrained Flow

In entrained flow systems, the fuel can be injected along with oxygen, a mixture of steam and oxygen, or air. There is an area where most of the molten slag is collected. The output products that reach high temperatures require prior cooling before the fuel gas goes through its cleaning process, thus resulting in a decrease in the system's thermal efficiency. This gasification technology produces little methane, uses a relatively compact equipment, and involves short reaction times due to high operating temperatures (1300–1800 K) (Larson et al. 2001). During the past decade, Texaco was the most widespread commercial application system of entrained bed gasification to produce synthesis gas. In addition to Texaco's Gasifier, other technologies, such as Shell's and Destec's, were initially developed on a commercial scale in the United States. Although these processes, generally speaking, have lower thermal efficiency than fixed-bed and fluidized bed systems, its resulting gas presents low tar, hydrocarbons that are heavier than methane and nitrogen compound concentrations. Shell, Destec, and Texaco processes can produce up to 2000 t/day (Sánchez et al. 2008).

On account of their proven performance, entrained bed gasifiers have been selected for demonstrations in BIG-GTCC systems in the United States and in a few European countries using coal as fuel. Among such reactors, the following can be described (Sánchez et al. 2008):

- Koopers-Totzek gasifier: Commercially, this is the most important entrained bed reactor in operation.

Figure 2.41 illustrates a Koppers-Totzek reactor. It works at atmospheric pressure and at higher temperatures than other gasifiers. There are two versions: two- and four-headed ones.

The first version of this reactor was developed in the 1951, which works horizontally and has an elliptical shape, while the second one was built in 1970 in India, showing improved performance. Each head contains two burners, and the reactor is coated with a water jacket and refractory material. The molten ashes are removed through the center of the gasifier.

- Texaco Gasifier: It was developed in the United States and consists of a pressurized vertical cylinder covered with a refractory material, which preferably operates with residual bituminous coals mixed with water. Figure 2.42 shows the process flow chart of this type of reactor. In commercial applications, this

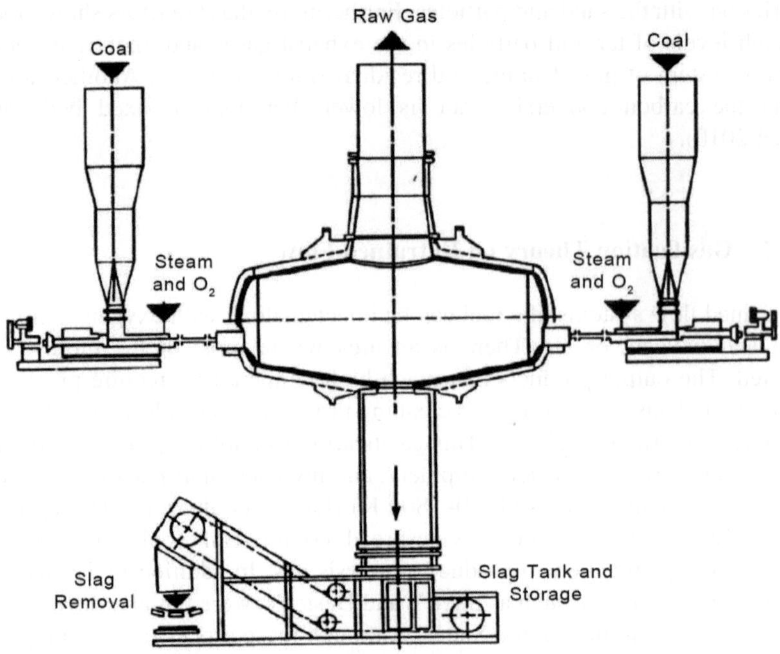

Fig. 2.41 Koppers-Totzek Gasifier (Adapted from Sánchez et al. 2008)

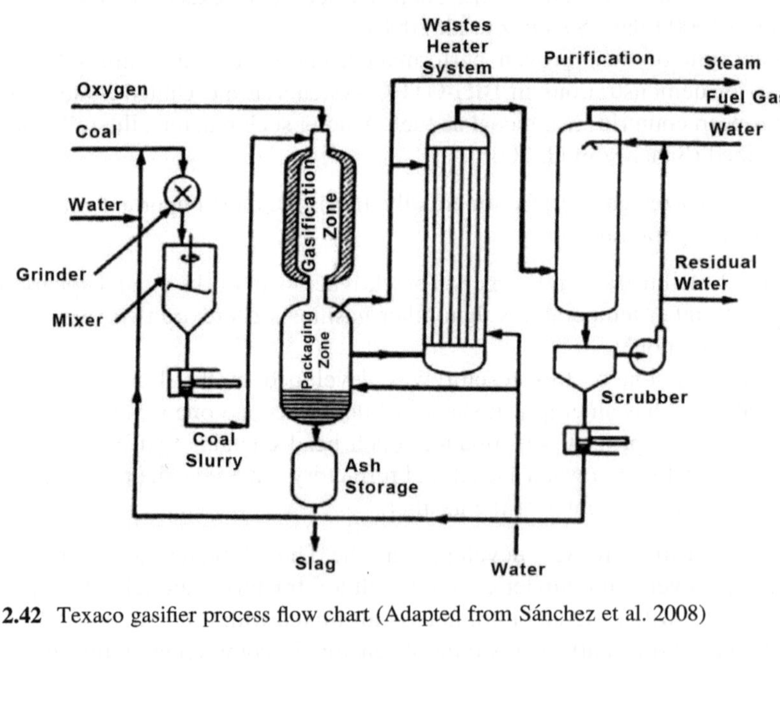

Fig. 2.42 Texaco gasifier process flow chart (Adapted from Sánchez et al. 2008)

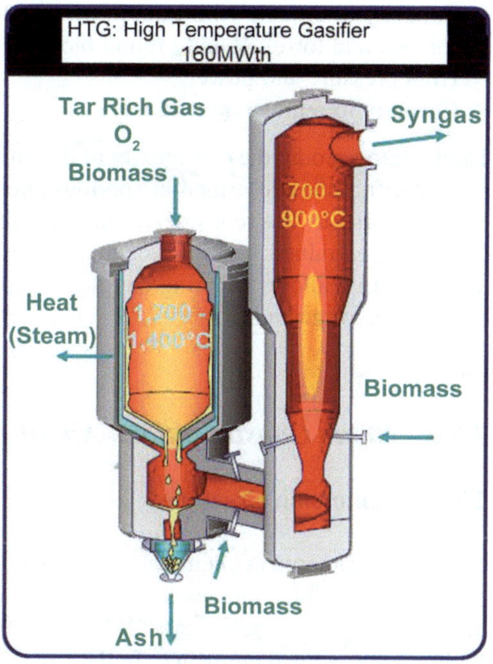

Fig. 2.43 Entrained flow gasifier. CHOREN Technology (Adapted from Worley and Yale 2012)

gasifier produces syngas to produce methanol and acetic anhydride. In California (USA), there is a plant with enough capacity to process 910 t/day of fuel.

- Otto-Rummel Gasifier: This equipment also works at atmospheric pressure and is divided into three stages. In the first stage, the coal is dried and pulverized, which is then injected along with steam and oxygen. In the second stage, the gasification reactions occur and, in the final stage, the resulting gas is collected and cooled. A reactor of this type with 1.8 m in diameter was installed on a commercial scale in 1960 in Germany, which operated steadily for 18 months.
- Shell-Koppers Gasifier: This reactor is a pressurized version of the Koppers-Totzek gasifier with very similar design, but with higher thermal efficiency. There are currently two plants with a capacity of 1000 t/day, one being in Germany and the other one in the Netherlands.
- CHOREN Gasifier: This gasification technology was developed by CHOREN Industries GmbH, one of the most efficient, reaching 81 % efficiency in cold gas (Worley and Yale 2012). The main fuel requirements of such technology are small ones (preferably <6 mm) and with low moisture content (Fig. 2.43).

2.9.1.8 Entrained Flow Gasification Advantages

- High quality of resulting gas: it has low tar and methane contents, due to high operating temperatures (above 1300 °C) and to high levels of CO and H_2

- Capacity to operate with liquid flows or particulate matters of low particle size (bio-oil and torrefied and ground biomass)
- High pressure and power
- Capacity to remove molten ashes

Taking into account the complexity of the physical properties of sugar cane bagasse and the difficulties described in previous attempts of gasifying this fuel in fluidized bed reactors, it was considered the possibility of gasifying the bagasse in a CHOREN entrained bed gasifier after being pretreated, by using a conditioning technique that allows adapting the physical properties of bagasse to those required by this type of gasifier.

2.9.2 Some Considerations About Ammonia Cracking

2.9.2.1 Ammonia Cracking

The use of ammonia as chemical compound (medium) for storing hydrogen that can be easily dissociated and used in fuel cells and power plants is not novel, but it has been ongoing for over 40 years (T-Raissi 2002). Recently, the concept of ammonia economy has gained eminence. In the early 1970s when the concept of "Hydrogen Energy Economy" was being widely debated, it was envisioned that ammonia (NH_3) would provide a perfect storage medium for hydrogen production.

As hydrogen, ammonia is carbon-free and can be produced from any energy resource. However, there are also some significant advantages in terms of storage and transport. Ammonia can be liquefied at room temperature and pressures of 8–10 bar, furthermore, it can be stored in a similar manner to propane, whereas hydrogen requires expensive cryogenic storage mechanisms (Cheddie 2012). Table 2.19 presents the basic physical properties of ammonia.

2.9.2.2 Hydrogen Generation from Ammonia

Hydrogen can be produced from ammonia using different methods. Most of the reported experiences are based on thermal decomposition or catalytically cracking of ammonia into nitrogen and hydrogen. Some papers also address the hydrolysis of ammonia products, as ammonia borane for such a purpose.

Catalytic Decomposition of Ammonia

$$2NH_3 + 92.4kJ \rightarrow N_2 + 3H_2 \tag{2.47}$$

Table 2.19 Physical properties of ammonia (Adapted from Thomas and Parks 2006)

Hydrogen content					
Hydrogen weight fraction	17.65	wt.%	Hydrogen volume density	0.105	kg/l
Solid phase					
Melting point	–78	°C	Latent heat of fusion (1 atm at triple point)	–337.37	kJ/kg
Liquid phase					
Vapor pressure (21 °C)	8.88	bar	Liquid density (1 atm at boiling point)	682	kg/m^3
Boiling point (at 1 atm)	–33.5	°C	Liquid/gas equivalent (1 atm and 15 °C)	947	vol/vol
Latent heat of vaporization (1 atm at boiling point)	1371.	kJ/kg			
Critical temperature	132.4	°C	Critical pressure	112.8	bar
Gas phase					
Gas density (1 atm at boiling point)	0.86	3	Gas density 15 °C, 1 atm	0.73	kg/m^3
Compressibility (Z) (1 atm at 15 °C)	0.9929	Specific gravity (air = 1) (1 atm at 20 °C)	0.597		
Specific volume 1 atm at 20 °C)	1.411	3	Viscosity	.000098	Poise
Heat capacity at constant pressure (Cp) (1 atm at 15 °C)	.037	kJ/(mol K)	Heat capacity at constant volume (Cv) (1 atm at 15 °C)	.028	kJ/(mol K)
Critical density:	0.24	g/mL	Entropy, gas at 25 °C,1 atm.	45.97	cal/mol°C
Thermal conductivity	22.19	mW/(m K)			
Miscellaneous					
Water solubility (1 atm at 0 °C)	862	vol/vol	Auto ignition temperature	630	°C
Lower flammable limit in air	15 %	By volume	Upper flammable limit in air	28 %	By volume
Molecular weight:	17.03				

Ammonia is unstable at high temperatures and starts decomposing at 200 °C (Cheddie 2012). The slightly endothermic decomposition reaction is shown in Eq. (2.47). Thermodynamically, 98–99 % conversion of ammonia to hydrogen is possible at temperatures that are as low as 425 °C. However, in practice, the conversion rate depends on temperature and catalysts being used. Thermal decomposition or catalytic cracking is the most common method of hydrogen generation from ammonia. For large scale hydrogen generation ($>1000\ m^3/h$), natural gas reforming remains the most cost-effective process, despite the fact that, on a small-scale generation, ($<10\ m^3/h$), ammonia cracking becomes slightly less costly than

Table 2.20 Life cycle cost analysis of hydrogen production via various processes (Lipman and Shah 2007a, b)

	Cost of H_2 production, US$/(m^3/h)			
Scale of H_2 production (m^3/h)	Water electrolysis	Natural gas reforming	Methanol reforming	Ammonia cracking
10	0.943	0.390	0.380	0.343
100	0.814	0.261	0.285	0.279
1000	0.739	0.186	0.226	0.241

natural gas reforming (see Table 2.20) (Lipman and Shah 2007a, b). This result is based on the lifecycle cost analysis, taking into account investment and operation costs.

Early studies conducted on ammonia decomposition are more heavily focused on ammonia synthesis and the so-called iron-based catalysts. Since then, various metals, alloys, and noble metal compounds have been tested for ammonia decomposition. These include Fe, Ni, Pt, Ru, Ir, Pd, Rh; alloys such as Ni/Pt, Ni/Ru, Pd/Pt/Ru/La; and alloys of Fe with other metal oxides, including Ce, Al, Si, Sr, and Zr (Thomas and Parks 2006). Various catalysts have been investigated for decomposing ammonia to produce hydrogen for alkaline fuel cells. These include WC, Ni/Al_2O_3, $NiCeO_2/Al_2O_3$, Cr_2O_3, Ru/ZrO_2, and Ru on carbon nanofibers. Cesium-promoted ruthenium supported on graphite was also found to be very promising (Cairns and Simons. 1968). For these catalysts, a minimum temperature of 300 °C is required for an efficient release of ammonia for hydrogen production.

The catalysts performance can be quantified by using the hydrogen production rate, conversion fraction of ammonia (fraction of ammonia that is converted to hydrogen), and activation energy. The hydrogen generation rate from ammonia decomposition was measured experimentally, typically in units of millimoles of hydrogen produced every minute per gram of catalyst being loaded (mmol/min/g). The performance of various catalysts for ammonia decomposition, reviewed in this section, is summarized in Table 2.21.

Table 2.21 shows that the best ammonia conversion and hydrogen generation rates via thermal decomposition are obtained using Ru/CNT catalysts treated with potassium-based alkalis.

Ni achieves very good results as well, but it requires higher temperatures (500–600 °C) to achieve performance which is equivalent to Ru at 400 °C. The advantage of Ni is that it is less expensive than Ru, and it can be loaded at high concentrations to obtain the desired results.

An anode supported SOFC (with anode thickness of 500 μm, 40 % porosity, and 50 % Ni in volume) requires Ni loading of 0.134 g/cm^2. If it is operated at a current density of 5000 mA/cm^2, it consumes hydrogen at 11.6 mmol/min/g of catalyst. If the cell operates at 600 °C, then Ni can safely decompose ammonia at the required rate.

Supports have also been focus of researchers. The support's function is to enhance dispersion, thus increasing the effective area of the active catalyst. The

Table 2.21 Summary of ammonia decomposition catalysts performance (Lipman and Shah 2007a, b)

Catalyst/support	Temp. (°C)	Rate of H_2 Gen. (mmol/min/g)	Conv. eff. (%)
Nano-sized Ni/Santa Barbara Amorphous (SBA)-15 support	450	8.4	25.0
	500	17.4	52.1
	550	26.8	80.1
	600	31.9	95.2
	650	33.2	99.2
Ni/SBA-15	550	12.7	37.8
Ni/SiO_2	400	0.4	1.4
	500	3.3	10.5
	550	6.8	21.6
	600	11.4	36.4
	650	21.1	70.0
Ni/SiO_2	550	11.6	34.6
Ni/Al_2O_3	550	12.7	37.8
Ni/Al_2O_3	500	24.1	71.9
Ni/Al_2O_3 coated cordierite monolith	550	16.5	50.0
Ni/Al_2O_3(unsupported particles <200 μm)		13.2	40.0
Ir/SiO_2	400	1.2	3.9
	500	5.7	18.2
	600	17.6	56.0
	700	30.6	98.0
Ru/SiO_2	400	4.5	14.3
	500	20.0	64
	600	30.3	97
	650	30.9	99
Ru/ZrO_2Ru/Al_2O_3	550	25.8	77.0
		23.5	73.7
Ru/CNT	400	6.2	3.7
Ru/K-CNT		12.2	7.3
Ru/K-ZrO_2-BD		8.5	5.3
Ru/ZrO_2		3.7	2.2
Ru/Al_2O_3		3.8	2.3
Ru/MgO		5.4	3.2
Ru/TiO_2		4.3	2.6
Ru/CNT Ru/MgO–CNT	400	6.0	9.0
		8.7	13.0
Ru/CNT treated with KNO_3	400	33.3	49.7
Ru/CNT treated with KOH		31.6	47.2
Ru/CNT treated with K_2CO_3		31.3	46.7

All studies were based on an ammonia gas hourly space velocity (GHSV) of 30.000 mL/h/g of catalyst, except for (a) GHSV = 150.000 mL/h/g and (b) GHSV = 60.000 ml/h/g

support should be stable under reaction conditions and have a high specific surface area. For the catalyst Ru, there are various supports, including silica, alumina, graphitized carbon, carbon nanotubes, and nitrogen doped carbon nanotubes (Cheddie 2012). Yin et al. (2004a) ranked the supports for Ru in a decreasing order of activity measured by the ammonia conversion rate: Carbon nanotube (CNT) > MgO > TiO_2 > Al_2O_3 > ZrO_2 > AC > ZrO_2/BD. It was proposed that CNTs have had the best performance because they allowed the best dispersion of Ru, and also because of their high purity. CNTs have the additional advantage of presenting high conductivity which aids in electron transfer, thus facilitating the recombinative nitrogen desorption phase. They further showed that the use of the MgO-CNT support resulted in better performance of the catalyst Ru than by using the MgO base or the CNT base alone (Yin et al. 2004b). Temperature-programmed hydrogenation results showed that MgO resulted in an even greater stability for the CNT.

2.9.2.3 Production of Hydrogen from Ammonia

A recent comparison of the economics of various methods of hydrogen production showed that the catalytic cracking of ammonia, steam reforming of natural gas and methanol are all more economical methods of hydrogen production than water electrolysis (Silversand 2002). Natural gas reforming is estimated to be the method which involves the lowest cost of production on larger scales, but on small scale production of ten cubic normal meters an hour, ammonia cracking is estimated to be the least costly method. Table 2.22 shows the estimated production costs for these methods as a function of the production rates of ammonia.

In summary, with the anhydrous ammonia, sustainable hydrogen production can be achieved to be used as a fuel source expanding the use of fuel cells in stationary and mobile power generating applications and reducing any inherent emissions of

Table 2.22 Economic comparison of ammonia cracking and other hydrogen production methods

Unit size	Electrolysis				Natural gas reforming			
(Nm^3/h)	I	R	M	T	I	R	M	T
10	\$0.286	\$0.64	\$0.014	**\$0.943**	\$0.286	\$0.09	\$0.014	**\$0.390**
100	\$0.157	\$0.64	\$0.014	**\$0.814**	\$0.157	\$0.09	\$0.014	**\$0.261**
1000	\$0.082	\$0.64	\$0.014	**\$0.739**	\$0.082	\$0.09	\$0.014	**\$0.186**
Unit size	Methanol reforming				Ammonia cracking			
(Nm^3/h)	I	R	M	T	I	R	M	T
10	\$0.214	\$0.15	\$0.014	**\$0.380**	\$0.143	\$0.19	\$0.014	**\$0.343**
100	\$0.119	\$0.15	\$0.014	**\$0.285**	\$0.079	\$0.19	\$0.014	**\$0.279**
1000	\$0.06	\$0.15	\$0.014	**\$0.226**	\$0.041	\$0.19	\$0.014	**\$0.241**

Source: Silversand (2002)

Note: The production costs include investment costs (I), operation costs (R), maintenance costs (M), and total costs (T). All figures were converted to U.S. dollars from the Swedish Krona, using the conversion rate of 6.99 Krona/dollar, and are in units of dollars per cubic normal meter an hour (Nm3/h) of hydrogen production

greenhouse gases. Depending on how the ammonia or ammonia complex is initially produced, full fuel-cycle emissions of greenhouse gases and other air pollutants can be moderate or low/zero, which makes a further exploration of these types of systems rather interesting.

References

Ahmed A, Abdel M, Farag MAS (2011) Roles of microalgae and bacteria in hydrogen production as one of the renewable energy resources. Bull Environ Res 12(2):153–173

Abreu AJ (2012) Desenvolvimento e caracterização de catalisadores de níquel suportados em matrizes CeO-ZrO_2-Al_2O_3, 151 f. Teses (Doutorado)—Curso de Curso de Físico-química, Usp, São Carlos

Aiche T (2005) Steam reforming of methane and bio-ethanol: post-graduate course bioenergy—theory and application. Department of Energy Technology, Fraunhofer Institute for Solar Energy Systems (ISE), Helsinki University of Technology—HUT, pp 18–19

Aizquierdo U, Barrio VL, Lago N, Requies J, Cambra JF, Guemez MB, Arias PL (2012) Biogas steam and oxidative reforming processes for synthesis gas and hydrogen production in conventional and microreactor reaction systems. Int J Hydrogen Energy 37:13829–13842

Ali T-Raissi (2002) Technoeconomic analysis of area II hydrogen production—Part II. Hydrogen from ammonia and ammonia-borane complex for fuel cell applications. Proceedings of the 2002 U.S. DOE Hydrogen Program Review. NREL/CP-610-32405

Alves SC (2005) Reforma a Vapor do Metano para a Produção de Hidrogênio: Estudo Termodinâmico e Protótipo de Modelo Matemático de Reator com Membrana. Dissertation, Uberlândia: Universidade Federal de Uberlândia

Andrade RV (2007) Gaseificação de biomassa: Uma análise teórica e experimental, 205 f. Teses (Doutorado em Engenharia Mecânica)—Instituto de Engenharia Mecânica, Universidade Federal de Itajubá, Itajubá

ANEEL (Brasil) (2008) Atlas de Energia Eletrica do Brasil: Agencia Nacional de Energia Eletrica, 3rd edn. ANEEL, Brasilia, 236 p

ANEEL BRASIL (2014) http://www.aneel.gov.br/aplicacoes/capacidadebrasil/operacaocapacidadebrasil.asp. Accessed 17 May 2014

Araki S, Hino N, Mori T, Hikazudani S (2010) Autothermal reforming of biogas over a monolithic catalyst. J Nat Gas Chem 19(5):477–481

Arizona State University. http://www.asulightworks.com/blog?page=2. Accessed 17 July 2013

Avraam DG, Halkides TI, Liguras DK, Bereketidou OA, Goula MA (2010) An experimental and theoretical approach for the biogas steam reforming reaction. Int J Hydrogen Energy 35:9818–9827

Azenha MB (2013) A energia hidrelétrica não é limpa, nem barata. Interview, http://www.viomundo.com.br/entrevistas/bermann-a-energia-hidreletrica-nao-e-limpa-nem-barata.html. Accessed 10 Mar 2013

Basso G, Farret FA, Gonzatti F, Ferrigolo FZ, Franchi D, Miotto M (2013) Projeto e dimensionamento de uma lanta a células a combustível para redução do consumo de energia nos horários de pico de demanda. In: 2° Fórum Internacional Ecoinovar, Santa Maria, RS

Basu P (2006) Combustion and gasification in fluidized beds. Taylor & Francis, Abingdon, pp 355–357

Bertuccioli L, Chan A, Hart D, Lehner F, Madden B, Standen E (2014) Study on development of water electrolysis in the EU. Final report. In: Fuel cells and hydrogen joint undertaking, 160p

Beurden PV (2004) On the catalytic aspects of steam-methane reforming: a literature survey. ECN-I-04-003

Bhandari R, Trudewind CA, Zapp P (2014) Life cycle assessment of hydrogen production via electrolysis—a review. J Cleaner Prod 85:151–163

Braga LB (2010) Análise econômica do uso de célula a combustível para acionamento de ônibus urbano, 99 f. Dissertation—Curso de Engenharia Mecânica, Departamento de Energía, UNESP, Guaratinguetá

Brenna G (2010) New catalyst for the H_2 production by water-gas shift reaction processes. Tese (Doutorado). Curso Química Industrial. Faculdade de Química Industrial. Universidade de Bologna

Brisse A, Schefold J, Zahid M (2008) High temperature water electrolysis in solid oxide cells. Int J Hydrogen Energy 33:5375–5382

Burgess SJ, Tamburic B, Zemichael F, Hellgardt K, Nixon PJ (2011) Solar driven hydrogen production in green algae. Adv Appl Microbiol 75:71–110

Cai X, Dong X, Lin W (2006) AutotheRMAL REFORMING OF METHANE over Ni catalysts supported on $CuO\text{-}ZrO_2\text{-}CeO_2\text{-}Al_2O_3$. J Nat Gas Chem 15(2):122–126

Cairns EJ, Simons EL (1968) Ammonia-oxygen fuel cell. Nature 217(5130):780–781

Casanovas ARM, Leitenburg C, Trovarelli A, Lorca J (2010) Ethanol steam reforming and water gas shift over Co/ZnO catalytic honeycombs doped with Fe, Ni, Cu, Cr and Na. Int J Hydrogen Energy 35:7690–7698

Cheddie D (2012) Ammonia as a hydrogen source for fuel cells: a review. In: Minic D (ed), Hydrogen energy—challenges and perspectives. InTech. ISBN: 978-953-51-0812-2

Chena Y, Wang Y, Xub H, Xiong G (2008) Efficient production of hydrogen from natural gas steam reforming in palladium membrane reactor. Appl Catal Environ 80:283–294

CIRCE (2005) Curso Técnico en Sistemas de Energías Renovables. Centro de Investigación de Recursos y Consumos Energéticos CIRCE, Centro Politécnico Superior, Universidad de Zaragoza, España

Claassen PAM, de Vrije T (2006) Non-thermal production of pure hydrogen from biomass: HYVOLUTION. Int J Hydrogen Energy 3:1416–1423

Cortés OEJ (2009) Biocombustíveis a partir de microalgas: modelagem e análise de fotobiorreatores. Tese (Doutorado)—Curso Energia e Ambiente, Centro Interdisciplinar de Energia e Ambiente (CIEnAm), Universidade Federal da Bahia

CRESESB (2012) http://www.cresesb.cepel.br/sundata/index.php. Accessed 15 August 2012

Dantas GA, Leite ALS (2013) Os custos da energia eólica brasileira. http://www.nuca.ie.ufrj.br/gesel/artigos/Os_custos_energia.pdf. Accessed 24 Feb 2013

Das D, Veziroglu TN (2008) Advances in biological hydrogen production processes. Int J Hydrogen Energy 33:6046–6057

Duane MB, Ariff GD, James BD, Lettow JS, Thomas CE, Kuhn RC (2002) Cost and performance comparison of stationary hydrogen fueling appliances: Task 2 report. In: Directed Technologies, Inc. www.directedtechnologies.com

Dutra R (2008) CRESESB: Energia Eólica: Princípios e Tecnologia, 58 p

Effendi A, Hellgardt K, Zhang ZG, Yoshida T (2005) Optimising H_2 production from model biogas via combined steam reforming and CO shift reactions. Fuel 84:869–874

Carnieletto R (2011) Aproveitamento de energia vertida turbinável para produção de hidrogênio e geração distribuída. Universidade Federal de Santa Maria, Santa Maria, p 155

Energy Options. http://energy-options.info/2011/03/25/lab-biofuel-now-to-go-into-the-market place/. Accessed 17 July 2013

FAPESP (2013) Revista Pesquisa (Org.). Reforma energética. http://revistapesquisa2.fapesp.br/?art=3029&bd=1&pg=5&lg. Accessed 10 Mar 2013

Foglia D, Wukovits W, Friedl A, Ljunggren M, Zacchi G, Urbaniec K, Markowski M (2011) Effects of feedstock on the process integration of biohydrogen production. Clean Technol Environ Policies J 13(4):547–558. doi:10.1007/s10098-011-0351-7

Furlan AL (2012) Análise técnica e econômica do uso do hidrogênio como meio armazenador de energia elétrica proveniente de fontes eólicas, 86 f. Tese (Doutorado)—Departamento de Engenharia Mecânica, Universidade Estadual de Campinas, Campinas

Gibson TL, Kelly NA (2009) Predicting efficiency of solar powered hydrogen generation. Int J Hydrogen Energy 35:900–911

Gibson TL, Kelly NA (2008) Optimization of solar powered hydrogen production using photovoltaic electrolysis devices. Int J Hydrogen Energy 33:5931–5940

Haag S, Burgard M, Ernst B (2007) Beneficial effects of the use of a nickel membrane reactor for the dry reforming of methane: comparison with thermodynamic predictions. J Catal 252:190–204

Hallenbeck PC, Benemann JR (2010) Biohidrogen: the microbiological production of hydrogen fuel. In: Doelle HW, Rokem S (eds) Biotechnologyin EOLSS encyclopedia: biotechnology, vol 7. ISBN: 978-1-84826-711-4

Haryanto A, Fernando SD, Filip To SD, Steele PH, Pordesimo L, Adhikari S (2011) High temperature water gas shift reaction over nickel catalysts for hydrogen production: effect of supports, GHSV, metal loading, and dopant materials. J Thermodyn Catal 2:106. doi:10.4172/2153-0645.1000106

Hemschemeier AC (2005) The Anaerobic life of the photosynthetic alga Chlamydomonas Reinhardtii. Tese (Doutorado), University of Bochum, Germany

HY Generation (2014) http://hy-generation.com/21prod_h2.html. Accessed 10 July 2014

Inoue T, Kumar SN, Kamachi T, Okura I (1999) Hydrogen evolution from glucose with the combination of Glucose dehydrogenase and Hydrogenase from A. eutrophus H16. Chem Lett 2:147–148

IRENA—International Renewable Energy Agency (2013) Summary for policy makers: renewable power generation costs: Abu Dhabi, United Arab Emirates, 12 p

ITAIPU. https://www.itaipu.gov.br/energia/energia-disponivel-anual. Accessed 10 Apr 2014

Ivy J (2004) Summary of electrolytic hydrogen production, U.S. In: National Renewable Energy Laboratory, Golden, USA

Koroneos C, Dompros A, Roumbas G, Moussiopoulos N (2004) Life cycle assessment of hydrogen fuel production processes. Int J Hydrogen Energy 29:1443–1450

Kothari R, Buddhi D, Sawhney RL (2008) Comparison of environmental and economic aspects of various hydrogen production methods. Renew Sustain Energy Rev 12:553–563

Krona (2012) http://www.krona.srv.br/display05.htm. Accessed 10 Oct 2012

Kroposki B, Levene J, Harrison K, Sen PK, Novachek F (2006) Electrolysis: information and opportunities for electric power utilities. In: National Renewable Energy Laboratory, NREL/TP-581-40605

Kuhn J, Kesler O (2015) Carbon deposition thresholds on nickel-based solid oxide fuel cell anodes I. Fuel utilization. J Power Sources 277:443–454

Kwietniewska E, Tys J, Krzeminska I, Kozief W (2012) Microalgae-cultivation and application of biomass as a source of energy: a review. Instytut Agrofizyki im. Bohdana Dobrzańskiego, Lublin

Larson ED, Williams RH, Regis M, Leal LV (2001) A review of biomass integrated gasifier/gas turbine combined cycle technology and its application in sugarcane industries with an analysis for Cuba. Energy Sustain Dev 5(1):54–76

Lehninger AL, Nelson DL, Cox MM (2010) Princípios de bioquímica, 2nd edn. Editora Savier. ISBN 85-7378-026-6

Levin DB, Pitt L, Love M (2004) Biohydrogen production: prospects and limitations to practical application. Int J Hydrogen Energy 29:173–185

Liguras DK, Kondarides DI, Verykios X (2003) Production of hydrogen for fuel cell by steam reforming of ethanol over supported noble metal catalyst. Appl Catal B Environ 43:345–354

Lipman T, Shah N (2007a). UC Berkeley Transportation Sustainability Research Center, UC Berkeley. http://www.escholarshiporg/uc/item/7z69v4wp

Lipman T, Shah N (2007b) Ammonia as an alternative energy storage medium for hydrogen fuel cells. Scientific and Technical Review for Near-Term Stationary Power Demonstration UCB-ITS-TSRC-RR-2007-5

Logan BE, Hamelers B, Rozendal R, Schröder U, Keller J, Freguia S, Aelterman P, Verstraete W, Rabaey K (2006) Microbial fuel cells: methodology and technology. Environ Sci Technol 40:5181–5192. doi:10.1021/es0605016

Lopez RA (2004) Célula combustível a hidrogênio: fonte de energia da nova era. Artliber, São Paulo, 182 p
Ma K, Schicho RN, Kelley RM, Adams MWW (1993) Hydrogenase of the hyperthermophile Pyrococcus furiosus is an elemental sulfur reductase or sulfhydrogenase: evidence for a sulfur-reducing hydrogenase ancestor. Proc Natl Acad Sci USA 90:5341–5344
Ma K, Zhou ZH, Adams MWW (1994) Hydrogen production from pyruvate by enzymes purified from the hyperthermophilic archaeon, Pyrococcus furiosus: a key role for NADPH. FEMS Microbiol Lett 122:245–250
Maggio G, Freni S, Cavallaro S (1998) Light alcohols/methane fuelled molten carbonate fuel cells: a comparative study. J Power Sources 74(1):17–23
Maia TA, Bellido JDA, Assaf EM, Assaf JM (2007) Produção de hidrogênio a partir da reforma a vapor de etanol utilizando catalisadores Cu/Ni/γ-Al_2O_3. Quim Nova 30:339–345
Ministry of Agriculture, Livestock and Food Supply—Secretariat for Production and Agroenergy (2006) Brazilian Agroenergy Plan, Embrapa Publishing House, 118p. http://www.agricultura.gov.br/pls/portal/docs/page/mapa/planos/pna_2006_2011/plano%20nacional%20de%20agroenergia%202006%20-%202011-%20ingles.pdf. Accessed 30 Jan 2011
Paisley MA, Litt RD, Creamer KS (1991) Gasification of refuse-derived fuel in a high throughput gasification system. Energy Biomass Waste 14:1991
Pelczar JRMJ, Chan ECS, Krieg NR (2008) Microbiology, 5th edn. Tata Mc Graw Hill, New York
Pérez NP, Machin EB, Pedroso DT, Antunes JS, Silveira JL (2014) Fluid-dynamic assessment of sugarcane bagasse to use as feedstock in bubbling fluidized bed gasifiers. Appl Therm Eng 73 (1):238–244
Pimentel TTBC (2012) O Enfrentamento político dos conflitos socioambientais decorrentes da implantação de usinas hidrelétricas. Dissertation, Curso Planejamento de Gestão Ambiental, Universidade Catolica de Brasilia
Piroonlerkgul P et al (2008) Selection of appropriate fuel processor for biogas-fuelled SOFC system. Chem Eng J 140:341–351
Raj NT, Goic SR (2011) A review of renewable energy based cogeneration technologies. Renew Sustain Energy Rev 15(8):3640–3648
Reed TB, Das A (1998) Handbook of biomass downdraft gasifier engine systems. The Biomass Energy Foundation Press, Golden
Roberts JJ (2012) Análise de Desempenho de um sistema Hibrido de Geração de energia solar-eólico-diesel considerando variações probabilisticas da carga e dos recursos renováveis, 151f. Dissertation. Curso Engenharia Mecânica. Faculdade de Engenharia de Guaratinguetá
Rozendal RA (2007) Hydrogen production through biocatalyzed electrolysis. PhD thesis Wageningen University, Wageningen, The Netherlands
Saebea D, Arpornwichanop A, Patcharavorachot Y, Assabumrungrat S (2011) Adsorption-membrane hybrid system for ethanol steam reforming: thermodynamic analysis. Int J Hydrogen Energy 36:12234–14428
Sahara Forest Project (2013) http://www.flickr.com/photos/bellona-foundation/sets/72157626842235875/. Accessed 18 July 2013
Sánchez CG, Silva E, Gomes EO (2008) Gaseificação. In: Biomassa para Energia, vol 1, 1 edn. Editora da UNICAMP, Campinas, pp 284–368
Santos JHT (2004) Avaliação de um sistema de aquecimento do substrato na digestão anaeróbia de dejetos suínos. Dissertation, Universidade Federal de Viçosa
Scholz M, Melin T, Wessling M (2013) Transforming biogas into biomethane using membrane technology. Renew Sustain Energy Rev 17:199–212
Silva EP (1991) Introdução à tecnologia e economia do hidrogênio. Unicamp, Campinas, p 204. ISBN: 85-268-0174-0
Silva ME (2005) Análise Termoquímica de Reformador de Etanol: Produção de Hidrogênio para Acionamento de uma Célula a Combustível do Tipo PEM de 1 kW. Guaratinguetá, 108p. Dissertation. Curso em Engenharia Mecânica Departamento de Energia, Faculdade de Engenharia, Campus de Guaratinguetá, Universidade Estadual Paulista

Silva ME (2010) Análise experimental da reforma a vapor de etanol: aspectos técnicos, econômicos e ecológicos. Tese (Doutorado)—Curso em Engenharia Mecânica. Faculdade de Engenharia de Guaratinguetá, Universidade Estadual Paulista, Guaratinguetá

Silveira JL (2012) Energia: crise e planejamento. http://www.comciencia.br/reportagens/energiaeletrica/energia13.htm. Accessed 2 Mar 2012

Silveira JL, Braga LB, Souza ACC, Antunes JS, Zanzi R (2009) The benefits of ethanol use for hydrogen production in urban transportation. Renew Sustain Energy Rev 13(9):2525–2534

Silversand F (2002) Catalytic heat exchangers for small-scale production of hydrogen—feasibility study. Svenskt Gastekniskt Center

Singh RN, Sharma S (2012) Development of suitable photobioreactor for algae production—a review. Renew Sustain Energy Rev 16:2347–2353

Smolinka T, Günther M, Garche J (2011) Status and development potential of water electrolysis for producing hydrogen from renewable sources. National Organisation Hydrogen and Fuel Cell Technology, Berlin

Sorensen B (2005) Hydrogen and fuel cell: emerging technologies and applications. Elsevier Academic Press, Amsterdam, 450 p

Souza MMVM (2005) Supported nickel catalysts for steam reforming of methane. In: 2nd Mercosur Congress on Chemical Engineering, 4th Mercosur Congress on Process Systems Engineering, Rio de Janeiro

Spath PL, Mann (2000) Life cycle assessment of a natural gas combined-cycle power generation system. NREL/TP-570-27715 Colorado: U.S. Department of Energy

Stein W, Edwards J, Hinkley J, Sattler C (2009) Natural gas: solar-thermal steam reforming. In: Encyclopedia of electrochemical power sources, pp 300–312

Sugai MH (2012) Modelagem matemática de coluna de gaseificação de fotobiorreatores tubulares para cultivo de microalgas. Dissertação em Engenharia Química, Setor de Tecnologia, Universidade Federal do Paraná, Curitiba

Tamburic B, Zemichael FW, Maitland GC, Hellgardt K (2011) Parameters affecting the growth and hydrogen production of the green algae Chlamydomonas reinhardtii. Int J Hydrogen Energy 36:7872–7876

Takeguchi T et al (2002) Study on steam reforming of CH4 and C2 hydrocarbons and carbon deposition on Ni-YSZ cermets. J Power Sources 112:588–595

Tao X, Qi F, Yin Y, Dai X (2008) CO_2 reforming of CH_4 by combination of thermal plasma and catalyst. Int J Hydrogen Energy 33:1262–1265

Thomas G, Parks G (2006) Potential roles of ammonia in a hydrogen economy: a study of issues related to the use ammonia for on-board vehicular hydrogen storage. U.S. Department of Energy, February. http://hydrogen.energy.gov/pdfs/nh3_paper.pdf

T-Raissi A, Block DL (2004) Hydrogen: automotive fuel of the future. IEEE Power Energy Mag 2 (6):40–45

Trane R, Dahl S, Skjøth-rasmussen MS, Jensen AD (2012) Catalytic steam reforming of bio-oil. Int J Hydrogen Energy 37:6447–6472

University of Nevada (2013) http://hrc.unlv.edu/renewable/biofuels/rd_Photobioreactor.html. Accessed 17 July 2013

Ursúa A, Gandía LM, Sanchis P (2012) Hydrogen production from water electrolysis: current status and future trends. In: Proceedings of The IEEE 100, New York, pp 410–426

Usui N, Ikenouchi M (1997) Biological CO_2 fixation and utilization project by RITE. 1. Highly-effective photobioreactor system. Energy Convers Manag 38:487–492

Van de Beld L (2001) Cleaning of hot producer gas in a catalytic, reverse flow reactor, Final report for: Novem (EWAB Programme, Report No. 9605) and European Commission (AIR Programme, AIR-CT93-1436)

Van der Meijden CM (2010) Development of the MILENA gasification technology for the production of Bio-SNG. PhD Thesis, ECN-B-10-016

Vasconcelos N (2006) Reforma a Vapor do Metano em Catalisadores à Base de Níquel Promovidos com Nióbia. Dissertation, Universidade Federal Fluminense, Niteroi

Vreugdenhil BJ, Van der Drift A, Van der Meijden CM (2009) Co-gasification of biomass and lignite in the indirect gasifier Milena. In: Proceeding of Pittsburgh coal conference, Pittsburgh, USA, 20–23 September

Wang M, Wang Z, Gong X, Guo Z (2014) The intensification technologies to water electrolysis for hydrogen production—a review. Renew Sustain Energy Rev 29:573–588

Wang SJ, Yin SF, Li L, Xu BQ (2004a) Ng CF, Au CT. Appl Catal B Environ 52:287

Wang Y, Chin YH, Rozmiarek RT, Johnson BR, Gao Y, Watson J, Tonkovich AYL, Vander DP (2004b) Highly active and stable Rh/MgO–Al_2O_3 catalysts for methane steam reforming. Catal Today 98:575–581

Williams RH (1996) Biomass gasifier gas turbine power generating technology. Biomass Bioenergy 10(1996):149–166

Wongchanapai S, Iwai H, Saito M, Yoshida H (2013) Performance evaluation of a direct-biogas solid oxide fuel cell-micro gas turbine (SOFC-MGT) hybrid combined heat and power (CHP) system. J Power Sources 223:9–17

Woodward J, Cordray KA, Edmonston RJ, Blanco-Rivera M, Mattingly SM, Evans BR (2000a) Enzymatic hydrogen production: conversion of renewable resources for energy production. Energy Fuels 14:197–201

Woodward J, Mattingly SM, Danson M, Hough D, Ward N, Adams M (1996) In vitro hydrogen production by glucose dehydrogenase and hydrogenase. Nat Biotechnol 14:872–874

Woodward J, Orr M (1998) Enzymatic conversion of sucrose to hydrogen. Biotechnol Prog 14:897–902

Woodward J, Orr M, Cordray K, Greenbaum E (2000b) Enzymatic production of biohydrogen. Nature 405:1014–1015

Worley M, Yale J (2012) Biomass gasification technology assessment, consolidated report 2012. National Renewable Energy Laboratory (NREL)-SR-5100-57085

Xu G, Chen X, Honda K, Zhang ZG (2004) Producing H_2-rich gas from simulated biogas and applying the gas to a 50W PEFC stack. National Institute of Advanced Industrial Science and Technology (AIST). http://www.interscience.wiley.com

Yin SF, Xu BQ, Wang SJ, Ng CF, Au CT (2004a) Catal Lett 96:113

Yin SF, Zhang QH, Xu BQ, Zhu WX, Ng CF, Au CT (2004b) J Catal 224:384

Zainal A (2010) Gasification of lignocellulosic biomass in fluidized beds for renewable energy development. Renew Sustain Energy Rev 114:2852–2862

Chapter 3
Thermodynamic Analysis of Hydrogen Production Processes

Lúcia Bollini Braga, Celso Eduardo Tuna, Fernando Henrique Mayworm de Araujo, Lucas Fachini Vane, Daniel Travieso Pedroso, and José Luz Silveira

3.1 Thermochemical Process

It was conducted a physical–chemical study based on the concepts of Gibbs free energy, equilibrium constant, and degree of advancement, with the purpose of determining optimal pressure and temperature conditions for the steam reforming processes. Afterwards, the catalytic conversion rate of reactants into products was addressed by considering the parameters that had already been obtained in the physical–chemical analysis. Then, the thermodynamic efficiency was calculated for each process: steam reforming processes of ethanol, natural gas, and biogas.

3.1.1 Physicochemical Analysis

It was conducted an individual study on each of the reactors of the steam reforming process (reformers and water-gas shift reactors), making it possible to identify what are the effects of pressure and temperature for both devices.

L.B. Braga • C.E. Tuna • F.H.M. de Araujo • L.F. Vane • D.T. Pedroso
Group of Optimization of Energy Systems—GOSE, College of Engineering of Guaratinguetá, Institute of Bioenergy Research—IPBEN, São Paulo State University—UNESP, Dr. Ariberto Pereira da Cunha Ave, 333. Pedregulho, Guaratinguetá 12516-410, São Paulo, Brazil

J.L. Silveira (✉)
Group of Optimization of Energy Systems—GOSE, College of Engineering of Guaratinguetá, Institute of Bioenergy Research—IPBEN, São Paulo State University—UNESP, Dr. Ariberto Pereira da Cunha Ave, 333. Pedregulho, Guaratinguetá 12516-410, São Paulo, Brazil
e-mail: joseluz@feg.unesp.br

J.L. Silveira (ed.), *Sustainable Hydrogen Production Processes*, Green Energy and Technology, DOI 10.1007/978-3-319-41616-8_3

3.1.1.1 Gibbs Free Energy (G)

Gibbs free energy is a thermodynamic function that allows predicting whether a process is spontaneous or not. The Gibbs free energy change (ΔG^o) depends exclusively on the initial and final state of the system. When the system is at constant pressure and temperature conditions and the substances involved are in their standard state, one can calculate the Gibbs free energy variation, as shown in Eq. (3.1) (Adamin and Almendra 2002).

$$\Delta G^o = \Delta H^o - T\Delta S^o \tag{3.1}$$

According to Adamin and Almendra (2002), if there is any phase change (or transformation) with one or more substances at a given temperature range, the values corresponding to entropy (ΔS) and enthalpy (ΔH) should be taken into account by using a generalized form of Gibbs free energy change, according to Eq. (3.2).

$$\Delta G^o = a + bT \cdot \log T + cT^2 + eT^{-1} + fT \tag{3.2}$$

where a, b, c, e, and f are set constants of the substance under study.

When $\Delta G < 0$, the reaction occurs directly and spontaneously, when $\Delta G > 0$, it occurs in reverse and spontaneous order and, when $\Delta G = 0$, it is in equilibrium (Adamin and Almendra 2002).

3.1.1.2 Equilibrium Constant (K)

When studying a chemical reaction, a very important state function is the Gibbs free energy change. This function must be known and calculated in advance. Thus, it can be found whether the reaction is thermodynamically possible or not, and to what extent it can occur. In other words, you must know the reaction extent, and how and to what degree it is going to evolve under certain conditions until it reaches a state of equilibrium ($\Delta G = 0$) (Adamin and Almendra 2002).

According to Adamin and Almendra (2002), the relation between the equilibrium constant and the Gibbs free energy, in the case of an ideal gas, is given by Eq. (3.3).

$$K = e^{-\Delta G/RT} \tag{3.3}$$

3.1.1.3 Degree of Advancement (α)

Castelan (1986) stated that the degree of advancement (α) indicates the degree of evolution of a reaction, and that it depends on factors such as pressure and temperature. It is an index that ranges from 0 (when there are no chemical reactions) to 1 (when all reactants are transformed into products).

3.1.2 Gibbs Free Energy Change as a Function of Temperature

In order to carry out a physicochemical analysis of the steam reforming processes of ethanol, natural gas, and biogas, the reactions to form methane, ethane, ethanol, carbon dioxide, carbon monoxide, and steam were considered with their respective Gibbs free energy values for different temperatures. The tables with the Gibbs free energy as a function of temperature are presented in Appendix 1.

From these data, it is developed a polynomial model, type $y(x) = ax^0 + bx + cx^2$, which adjusts the curves for each formation reaction using the software Origin Pro 8.0., so as to determine constants A, B, and C of equation $\Delta G = A + BT + CT^2$, as it can be noted in Figs. 3.1, 3.2, 3.3, 3.4, 3.5, and 3.6.

It is observed the polynomial fit and values of A, B, and C for each reaction. These reaction formation equations were obtained because, through them, the steam reforming equations of each studied component and the *shift* reaction can be written.

For biogas, it is considered its composition from cattle manure. The analysis of this component was based on the biogas obtained (after the purification process) from a biodigestor prototype built by LOSE (Laboratory of Optimization of Energy Systems) with 75.7 % CH_4 and 24.3 % CO_2. The steam reforming of biogas, as it appears in literature, can be considered as being: the steam reforming of methane + dry reforming of methane. For the present study, these reactions were considered before the biogas went through the shift reactor, as it is presented in Eqs. (3.4) and (3.5).

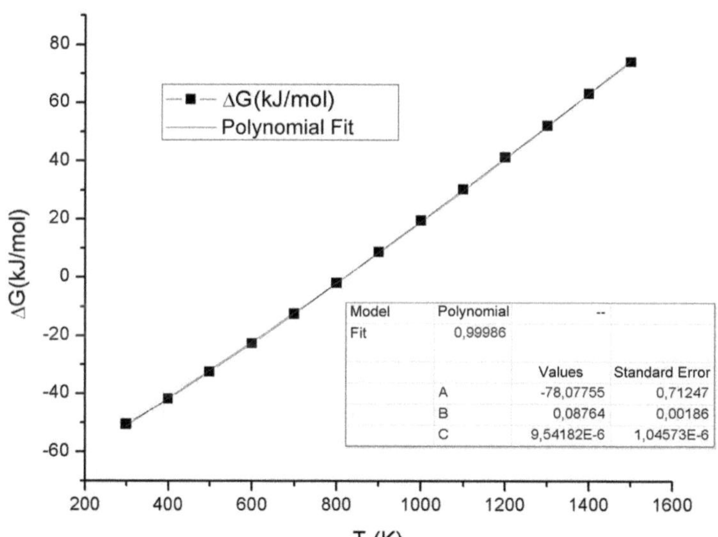

Fig. 3.1 Gibbs free energy values as a function of temperature—methane

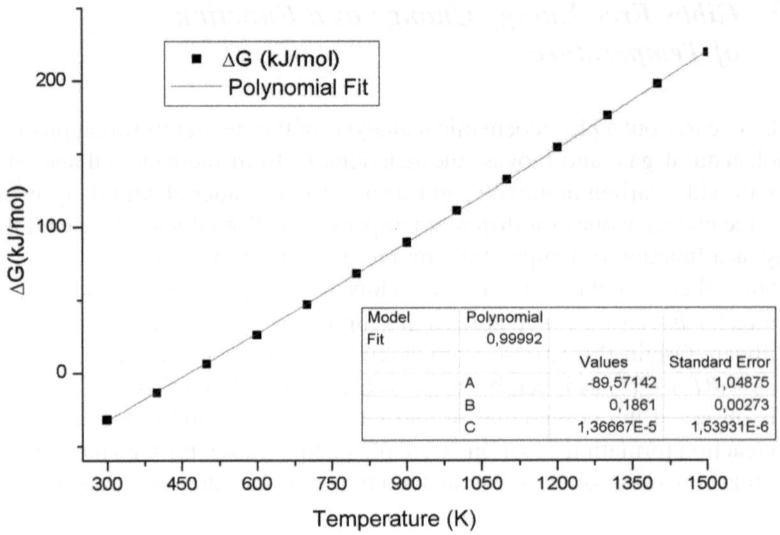

Fig. 3.2 Gibbs free energy values as a function of temperature—ethane

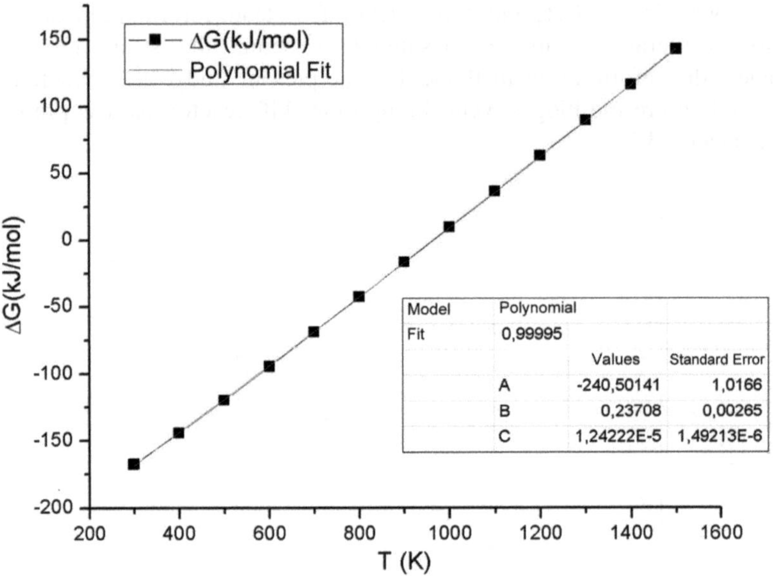

Fig. 3.3 Gibbs free energy values as a function of temperature—ethanol

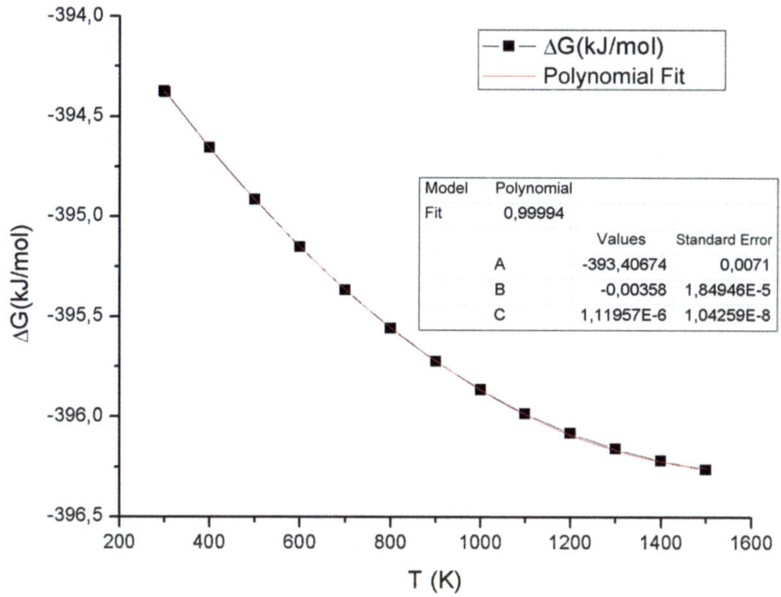

Fig. 3.4 Gibbs free energy values as a function of temperature—carbon dioxide

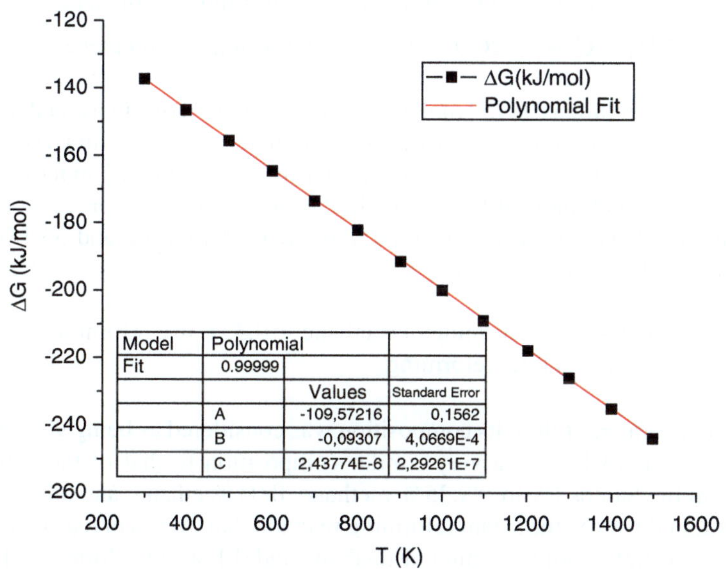

Fig. 3.5 Gibbs free energy values as a function of temperature—carbon monoxide

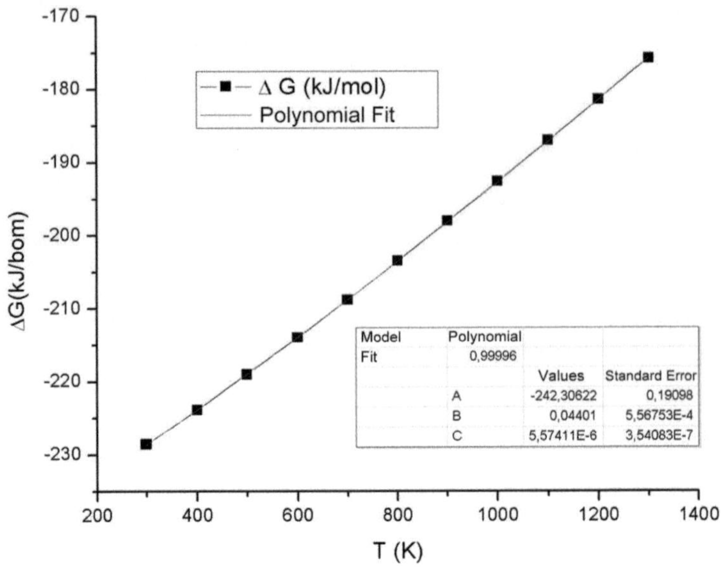

Fig. 3.6 Gibbs free energy values as a function of temperature—water

$$CH_4 + H_2O \rightarrow CO + 3H_2 \text{ (steam reforming of methane)} \quad (3.4)$$

$$CH_4 + CO_2 \rightarrow 2CO + 2H_2 \text{ (dry reforming of methane)} \quad (3.5)$$

Given 1 mol of biogas (0.757 mol of CH_4 and 0.243 mol of CO_2) and that all CO_2 reacts in the dry reforming with 1:1 CH_4 stoichiometry, 0.088 mol of CH_4 also react in the dry reforming. Therefore, when considering a total of 0.757 mol of CH_4, it is concluded that 0.669 mol (of CH_4) react in the steam reforming process. Thus, the final reaction of biogas consists of 66.9 % of steam reforming and 33.1 % of dry reforming, as shown in Eq. (3.6).

$$\begin{aligned} &0.669 \text{ steam reforming of methane} + 0.331 \text{ dry reforming} \\ &\quad = \text{biogas final reforming} \end{aligned} \quad (3.6)$$

For the natural gas (NG), its composition was considered as being 90 % methane and 10 % ethane, which is a very similar composition to that of the natural gas obtained in the Santos Basin (89.35 % methane, 8.03 % ethane, and others) (Krona 2012) and, in this way, the steam reforming reaction that occurs in reactor 1, for that component, is 0.9 steam reforming of methane and 0.1 steam reforming of ethane.

3.1.2.1 Gibbs Free Energy for Each Reforming Process (Reactor 1) and Shift Reaction (Reactor 2)

The chemical equations for the physicochemical analysis of the steam reforming and shift reactions are presented in Tables 3.1, 3.2, 3.3, and 3.4.

Table 3.1 Reactions for calculating the Gibbs free energy of the partial steam reforming of ethanol

Reactions	ΔG (kJ/mol) $= A + BT + CT^2$
$2C + 3H_2 + 1/2O_2 \rightarrow C_2H_5OH$	$-240.501 + 0.23708 \cdot T + 1.2422 \cdot 10^{-5} \cdot T^2$
+	+
$H_2 + 1/2O_2 \rightarrow H_2O$	$-242.306 + 0.04401 \cdot T + 5.5741 \cdot 10^{-6} \cdot T^2$
=	=
$2C + 4H_2 + O_2 \rightarrow C_2H_5OH + H_2O$	$-482.808 + 0.28109 \cdot T + 1.7996 \cdot 10^{-5} \cdot T^2$
−	−
$2C + O_2 \rightarrow 2CO$	$-219.144 - 0.18614.T + 4.87548 \cdot 10^{-6} \cdot T^2$
=	=
$4H_2 + 2CO \rightarrow C_2H_5OH + H_2O$	$-263.663 + 0.46723 \cdot T + 1.31208 \cdot 10^{-5} \cdot T^2$
Inverting the reaction	**Inverting the values**
$C_2H_5OH + H_2O \rightarrow 4H_2 + 2CO$	$263.663 - 0.46723 \cdot T - 1.31208 \cdot 10^{-5} \cdot T^2$

Table 3.2 Reactions for calculating the Gibbs free energy of the partial steam reforming of natural gas

Reactions	ΔG (kJ/mol) $= A + BT + CT^2$
Steam reforming of methane	
$C + 2H_2 \rightarrow CH_4$	$-78.0776 + 0.0876 \cdot T + 9.54182 \cdot 10^{-6} \cdot T^2$
+	+
$H_2 + 1/2O_2 \rightarrow H_2O$	$-242.306 + 0.04401 \cdot T + 5.5741 \cdot 10^{-6} \cdot T^2$
=	=
$C + 3H_2 + 1/2O_2 \rightarrow CH_4 + H_2O$	$-320.384 + 0.13165 \cdot T + 1.51159 \cdot 10^{-5} \cdot T^2$
−	−
$C + 1/2O_2 \rightarrow CO$	$-109.572 - 0.09307 \cdot T + 2.43774 \cdot 10^{-6} \cdot T^2$
=	=
$3H_2 + CO \rightarrow CH_4 + H_2O$	$-210.812 + 0.22472 \cdot T + 1.26782 \cdot 10^{-5} \cdot T^2$
Inverting the reaction	**Inverting the values**
$CH_4 + H_2O \rightarrow 3H_2 + CO$	$210.812 - 0.22472 \cdot T - 1.26782 \cdot 10^{-5} \cdot T^2$
Steam reforming of ethane	
$2C + 3H_2 \rightarrow C_2H_6$	$-89.5714 + 0.1861 \cdot T + 1.3666710^{-5} \cdot T^2$
+	+
$2H_2 + O_2 \rightarrow 2H_2O$	$-484.612 - 0.08802 \cdot T + 1.1148210^{-5} \cdot T^2$
=	=
$2C + 5H_2 + O_2 \rightarrow C_2H_6 + 2H_2O$	$-574.184 + 0.27412 + 2.48149 \cdot 10^{-5} \cdot T^2$
−	−
$2C + O_2 \rightarrow 2CO$	$-219.14432 - 0.18614 \cdot T + 4.87548 \cdot 10^{-6} \cdot T^2$
=	=
$5H_2 + 2CO \rightarrow C_2H_6 + 2H_2O$	$-355.04 + 0.46026 \cdot T + 1.99394 \cdot 10^{-5} \cdot T^2$
Inverting the reaction	**Inverting the values**
$C_2H_6 + 2H_2O \rightarrow 5H_2 + 2CO$	$355.04 - 0.46026 \cdot T - 1.99394 \cdot 10^{-5} \cdot T^2$
Steam reforming of natural gas	
$\Delta G_{\text{biogas}} = 0.9 \cdot (\Delta G_{\text{steam reforming of methane}}) + 0.1 \cdot (\Delta G_{\text{biogas steam reforming of ethane}})$	

Table 3.3 Reactions for calculating the Gibbs free energy of the partial steam reforming of biogas: dry and steam reforming of biogas

Reactions	ΔG (kJ/mol) $= A + BT + CT^2$
Steam reforming of methane	
$C + 2H_2 \rightarrow CH_4$	$-78.0776 + 0.0876 \cdot T + 9.54182 \cdot 10^{-6} \cdot T^2$
+	+
$H_2 + 1/2O_2 \rightarrow H_2O$	$-242.306 + 0.04401 \cdot T + 5.5741 \cdot 10^{-6} \cdot T^2$
=	=
$C + 3H_2 + 1/2O_2 \rightarrow CH_4 + H_2O$	$-320.384 + 0.13165 \cdot T + 1.51159 \cdot 10^{-5} \cdot T^2$
−	−
$C + 1/2O_2 \rightarrow CO$	$-109.572 - 0.09307 \cdot T + 2.43774 \cdot 10^{-6} \cdot T^2$
=	=
$3H_2 + CO \rightarrow CH_4 + H_2O$	$-210.812 + 0.22472 \cdot T + 1.26782 \cdot 10^{-5} \cdot T^2$
Inverting the reaction	**Inverting the values**
$CH_4 + H_2O \rightarrow 3H_2 + CO$	$210.812 - 0.22472 \cdot T - 1.26782 \cdot 10^{-5} \cdot T^2$
Steam reforming of methane	
$C + 2H_2 \rightarrow CH_4$	$-78.0776 + 0.0876 \cdot T + 9.54182 \cdot 10^{-6} \cdot T^2$
+	+
$C + O_2 \rightarrow CO_2$	$-393.4067 - 0.00358 \cdot T + 1.1195710^{-6} \cdot T^2$
=	=
$2C + 2H_2 + O_2 \rightarrow CH_4 + CO_2$	$-471.48429 + 0.08496 + 1.06614 \cdot 10^{-5} \cdot T^2$
−	−
$2C + O_2 \rightarrow 2CO$	$-219.14432 - 0.18614 \cdot T + 4.87548 \cdot 10^{-6} \cdot T^2$
=	=
$2H_2 + 2CO \rightarrow CH_4 + CO_2$	$-252.33997 + 0.2702 \cdot T + 5.78591 \cdot 10^{-5} \cdot T^2$
Inverting the reaction	**Inverting the values**
$CH_4 + CO_2 \rightarrow 2H_2 + CO$	$252.33997 - 0.2702 \cdot T - 5.78591 \cdot 10^{-5} \cdot T^2$
Steam reforming of biogas	
$\Delta G_{biogas} = 0.669 \cdot (\Delta G_{steam\ reforming}) + 0.331 \cdot (\Delta G_{biogas\ dry\ reforming})$	

Table 3.4 Reactions for calculating the *shift* reaction

Reactions	ΔG (kJ/mol) $= A + BT + CT^2$
$C + 1/2O_2 \rightarrow CO$	$-109.572 - 0.0937 \cdot T + 2.437710^{-6} \cdot T^2$
−	−
$C + O_2 \rightarrow CO_2$	$-393.407 - 0.00358 \cdot T + 1.11957 \cdot 10^{-6} \cdot T^2$
=	=
$CO_2 \rightarrow CO + 1/2O_2$	$283.8346 - 0.08949 \cdot T + 1.31817 \cdot 10^{-6} \cdot T^2$
+	−
$H_2 + 1/2O_2 \rightarrow H_2O$	$-242.306 + 0.04401 \cdot T + 5.57411 \cdot 10^{-6} \cdot T^2$
=	=
$H_2 + CO_2 \rightarrow CO + H_2O$	$41.52838 - 0.04548 \cdot T + 6.89228 \cdot 10^{-6} \cdot T^2$
Inverting the reaction	**Inverting the values**
$CO + H_2O \rightarrow H_2 + CO_2$	$-41.52838 + 0.04548 \cdot T - 6.89228 \cdot 10^{-6} \cdot T^2$

In each table, it was obtained the equation for each reforming process with their respective equations of Gibbs free energy as a function of temperature. In the case of biogas, the steam and dry reforming of methane was considered and, in the case of natural gas, the steam reforming of methane and ethane.

From the obtained values, it was developed graphs of Gibbs free energy as a function of temperature, as presented in Figs. 3.7, 3.8, 3.9, and 3.10.

As for the steam reforming of ethanol presented in Fig. 3.7, it is noted that the line intersects the abscissa axis at relatively high temperatures ($\Delta G < 0$), approximately 230 °C, indicating that it is an endothermic reaction.

In Fig. 3.8, the line of the steam reforming of natural gas intersects the abscissa axis at high temperatures ($\Delta G < 0$) of approximately 450 °C, indicating that it is an

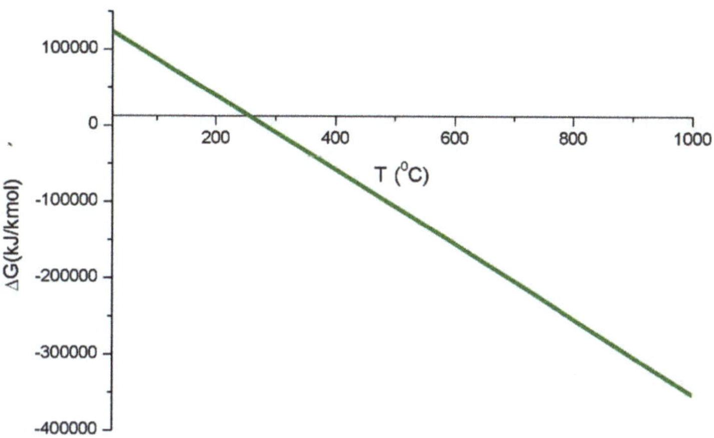

Fig. 3.7 Gibbs free energy as a function of temperature for the steam reforming process of ethanol

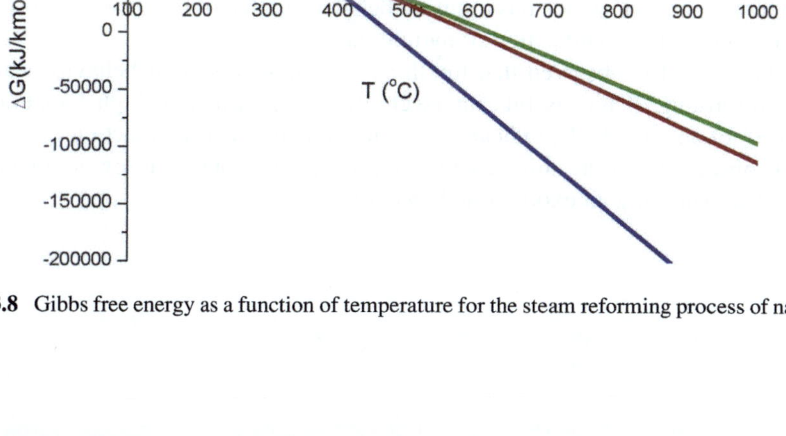

Fig. 3.8 Gibbs free energy as a function of temperature for the steam reforming process of natural gas

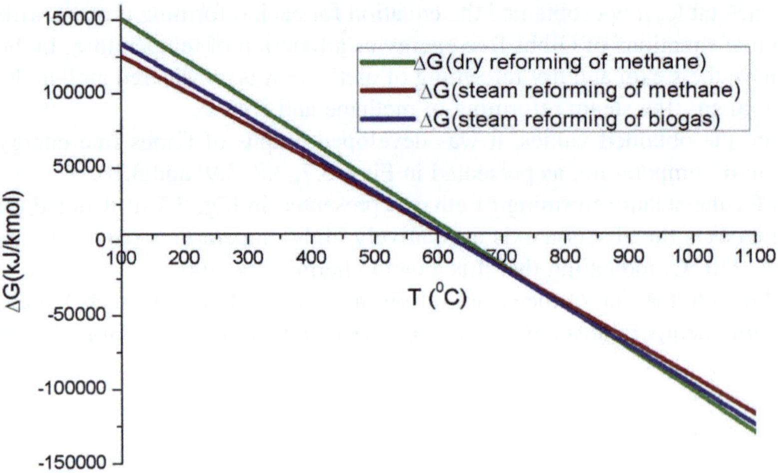

Fig. 3.9 Gibbs free energy as a function of temperature for the steam reforming process of methane and dry reforming of methane and biogas

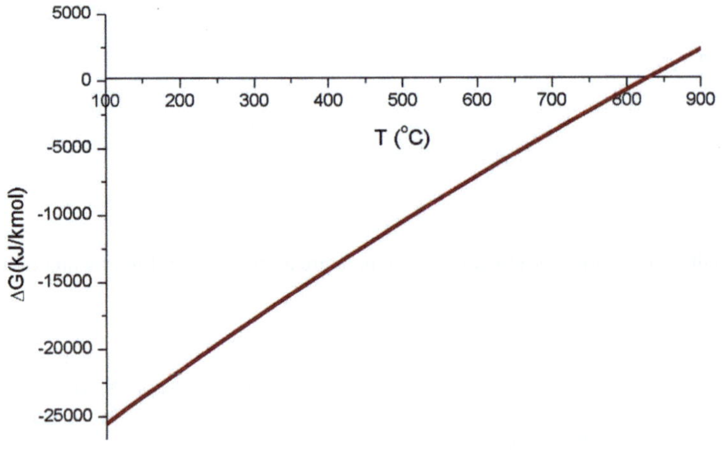

Fig. 3.10 Gibbs free energy as a function of temperature for the *shift* reactor process

endothermic reaction which has a similar behavior to that of the steam reforming of methane due to the composition of natural gas.

In Fig. 3.9, it is observed that the three lines have a similar behavior, and the steam reforming of biogas line intersects the abscissa axis at high temperatures ($\Delta G < 0$) of about 600 °C, indicating that it is an endothermic reaction.

It is observed that the shift reaction occurs spontaneously, at low temperatures ($\Delta G < 0$), exhibiting an exothermic behavior.

3.1.2.2 Equilibrium Constant

For calculating the equilibrium constant processes as a function of temperature, it is considered the following: $K = e^{-\Delta G/RT}$. The graphs plotted from the Gibbs free energy equations are shown in Figs. 3.11, 3.12, 3.13, and 3.14.

For the steam reforming of ethanol (Fig. 3.11), it is noted that the temperature and equilibrium constant are directly proportional. Once knowing that this constant is a quotient between products and reactants, it is found that, as the temperature increases, the rate of conversion also does, hence confirming that it is an endothermic process.

Figures 3.12 and 3.13 depict a similar behavior to that of the steam reforming of ethanol, as they are endothermic processes as well.

In the (water-gas) *shift* reaction process, shown in Fig. 3.14, it was observed that the equilibrium constant and temperature are inversely proportional. Therefore, as the temperature increases, the conversion of reactants into products decreases, thus confirming that it is an exothermic reaction.

3.1.2.3 Degree of Advancement

The degree of advancement was calculated for the equilibrium state. Thus, the equilibrium compositions of the shift reaction and reforming processes as a function of temperature and pressure were determined by establishing the chemical equilibrium of the reactions, as shown in Tables 3.5, 3.6, 3.7, and 3.8a, b, where (a) is the number of moles of reactants and products at the beginning and equilibrium of the reaction, and (b) indicates the total number of moles of the reaction in equilibrium and the atomic ratios of each component.

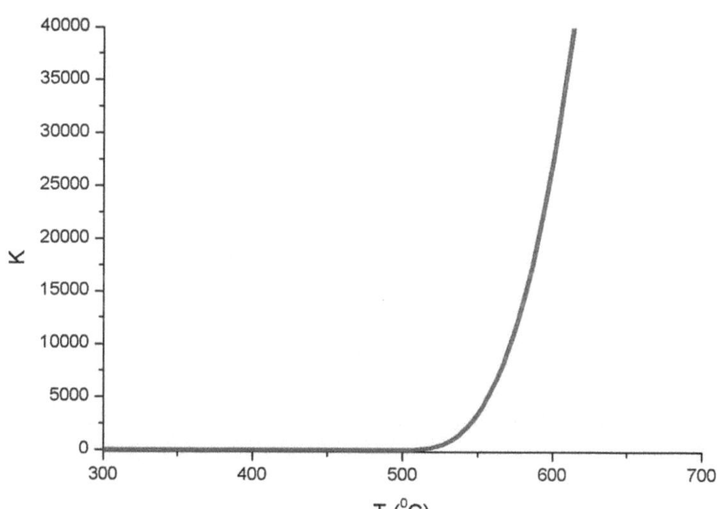

Fig. 3.11 Equilibrium constant of the steam reforming of ethanol as a function of temperature

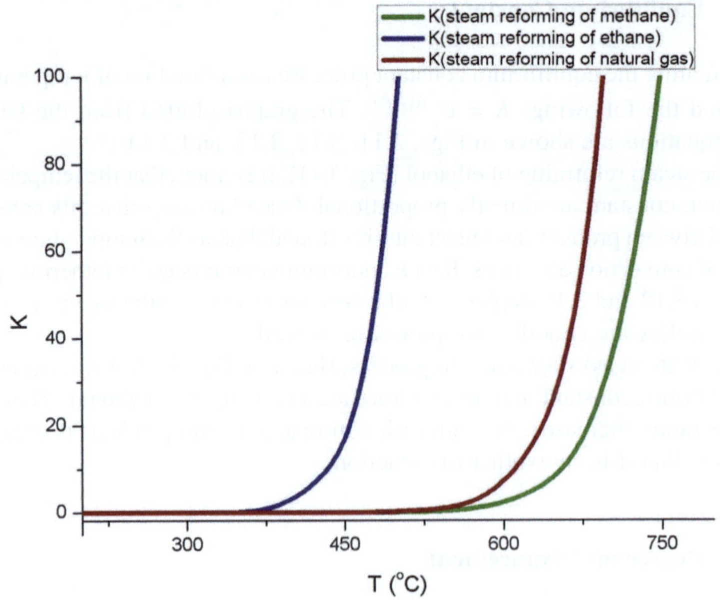

Fig. 3.12 Equilibrium constant of the steam reforming of natural gas as a function of temperature

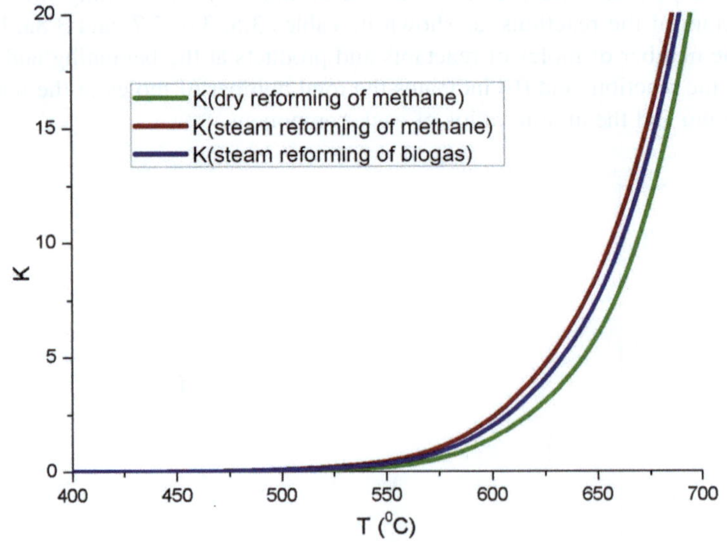

Fig. 3.13 Equilibrium constant of the steam reforming of biogas as a function of temperature

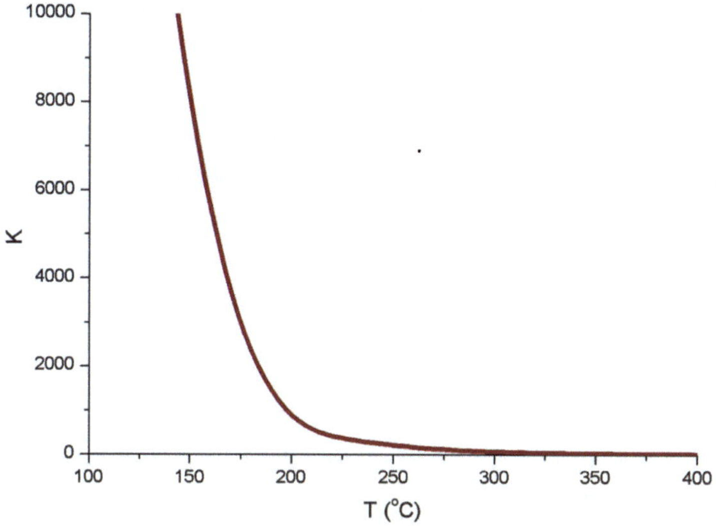

Fig. 3.14 Equilibrium constant of the (water-gas) *shift* reaction process

Table 3.5 Chemical equilibrium of the steam reforming reaction of ethanol

(a) Number of moles of reactants and products				
	$C_2H_5OH + H_2O \leftrightarrow 2CO + 4H_2$			
Beginning:	n	n	0	0
Moles:	$n^0 = n + n = 2n$			
Equilibrium:				
Moles:	$n(1-\alpha)$	$n(1-\alpha)$	$2n\alpha$	$4n\alpha$
(b) Total number of moles of the reaction and atomic ratios				
Total moles:	$n_{\mathrm{EQUILIBRIUM}} = n(1-\alpha) + n(1-\alpha) + 2n\alpha + 4n\alpha = 2n(1-\alpha) + 6n\alpha = 2n + 4n\alpha$			
	$n_{\mathrm{EQUILIBRIUM}} = 2n(1+2\alpha)$			
Atomic ratio:	$(1-\alpha)/2(1+2\alpha)$	$(1-\alpha)/2(1+2\alpha)$	$\alpha/(1+2\alpha)$	$2\alpha/(1+\alpha)$

The values calculated in Tables 3.5, 3.6, 3.7, and 3.8a, b were used to write the equilibrium constant and degree of advancement from Eqs. (3.7)–(3.23). In order to isolate the degree of advancement of the steam reforming reaction of ethanol, it was used the software Wolfram Mathematica (version 6.0.0), due to the complexity of the equation.

In the case of ethane, as the degree of advancement is raised to the seventh power and no solution has been found through the software, it was considered two degree of advancement equations: one for high temperatures, in this case considering the degree of advance as being close to 1, and another one for low temperatures considering the degree of advance as near zero (Sosa 2012).

The equilibrium constant for the steam reforming reaction of ethanol is presented in Eq. (3.7) and the degree of advancement in Eq. (3.8).

Table 3.6 Chemical equilibrium of the steam reforming reaction of natural gas

(a) Number of moles of reactants and products								
$CH_4 + H_2O \leftrightarrow CO + 3H_2$					$C_2H_6 + 2H_2O \leftrightarrow 2CO + 5H_2$			
Beginning:	n	n	0	0	n	$2n$	0	0
Moles:	$n^0 = n + n = 2n$				$n^0 = n + n = 2n$			
Equilibrium:					Equilibrium:			
Moles:	$n(1-\alpha)$	$n(1-\alpha)$	$n\alpha$	$3n\alpha$	$n(1-\alpha)$	$2n(1-\alpha)$	$2n\alpha$	$5n\alpha$

(b) Total number of moles of the reaction and atomic ratios				
	Methane reforming			
Total moles:	$n_{EQUILIBRIUM} = n(1-\alpha) + n(1-\alpha) + n\alpha + 3n\alpha = 2n(1-\alpha) + 4n\alpha = 2n + 2n\alpha$			
	$n_{EQUILIBRIUM} = 2n(1+\alpha)$			
Atomic ratio:	$(1-\alpha)/2(1+\alpha)$	$(1-\alpha)/2(1+\alpha)$	$\alpha/2(1+\alpha)$	$3\alpha/2(1+\alpha)$
	Ethane reforming			
Total moles:	$n_{EQUILIBRIUM} = n(1-\alpha) + 2n(1-\alpha) + 2n\alpha + 5n\alpha = 3n(1-\alpha) + 7n\alpha = 3n + 4n\alpha$			
	$n_{EQUILIBRIUM} = n(3+4\alpha)$			
Atomic ratio:	$(1-\alpha)/(3+4\alpha)$	$2(1-\alpha)/(3+4\alpha)$	$2\alpha/(3+4\alpha)$	$5\alpha/(3+4\alpha)$

Table 3.7 Chemical equilibrium of the steam reforming reaction and dry reforming of methane for calculating the biogas reforming reaction

(a) Number of moles of reactants and products								
$CH_4 + H_2O \leftrightarrow CO + 3H_2$					$CH_4 + CO_2 \leftrightarrow 2CO + 2H_2$			
Beginning:	n	n	0	0	n	n	0	0
Moles:	$n^0 = n + n = 2n$				$n^0 = n + n = 2n$			
Equilibrium:					Equilibrium:			
Moles:	$n(1-\alpha)$	$n(1-\alpha)$	$n\alpha$	$3n\alpha$	$n(1-\alpha)$	$n(1-\alpha)$	$2n\alpha$	$2n\alpha$

(b) Total number of moles of the reaction and atomic ratios				
	Methane reforming			
Total moles:	$n_{EQUILIBRIUM} = n(1-\alpha) + n(1-\alpha) + n\alpha + 3n\alpha = 2n(1-\alpha) + 4n\alpha = 2n + 2n\alpha$			
	$n_{EQUILIBRIUM} = 2n(1+\alpha)$			
Atomic ratio:	$(1-\alpha)/2(1+\alpha)$	$(1-\alpha)/2(1+\alpha)$	$\alpha/2(1+\alpha)$	$3\alpha/2(1+\alpha)$
	Dry reforming			
Total moles:	$n_{EQUILIBRIUM} = n(1-\alpha) + n(1-\alpha) + 2n\alpha + 2n\alpha = 2n(1-\alpha) + 4n\alpha = 2n + 2n\alpha$			
	$n_{EQUILIBRIUM} = 2n(1+\alpha)$			
Atomic ratio:	$(1-\alpha)/2(1+\alpha)$	$(1-\alpha)/2(1+\alpha)$	$\alpha/(1+\alpha)$	$\alpha/(1+\alpha)$

$$K = \frac{4\alpha^6}{(1+2\alpha)^4(1-\alpha)^2} P^4 \tag{3.7}$$

Table 3.8 Chemical equilibrium of the (water-gas) *shift* reaction

(a) Number of moles of reactants and products				
$CO + H_2O \leftrightarrow CO_2 + 4H_2$				
Beginning:	n	n	0	0
Moles:	$n^0 = n + n = 2n$			
Equilibrium:				
Moles:	$n(1-\alpha)$	$n(1-\alpha)$	$n\alpha$	$n\alpha$
(b) Total number of moles of the reaction and atomic ratios				
Total moles:	$n_{EQUILIBRIUM} = n(1-\alpha) + n(1-\alpha) + n\alpha + n\alpha = 2n(1-\alpha) + 2n\alpha = 2n$			
	$n_{EQUILIBRIUM} = 2n$			
Atomic ratio:	$(1-\alpha)/2$	$(1-\alpha)/2$	$\alpha/2$	$\alpha/2$

$$\alpha = \frac{\left(2 \cdot K + \sqrt{KP^3}\right)}{(2^{\frac{1}{3}}(2\sqrt{k} + P^3)\left(4K^{\frac{3}{2}} + 4KP^3 + \sqrt{k}P^6 + \sqrt{8K^{\frac{5}{2}}P^3 + 12K^2P^6 + 6K^{\frac{3}{2}}P^9 + KP^{12}}\right)^{\frac{1}{3}}} + \frac{\left(\frac{\left(2 \cdot K + \sqrt{KP^3}\right)}{\left(4K^{\frac{3}{2}} + 4KP^3 + \sqrt{k}P^6 + \sqrt{8K^{\frac{5}{2}}P^3 + 12K^2P^6 + 6K^{\frac{3}{2}}P^9 + KP^{12}}\right)^{\frac{1}{3}}}\right)}{2^{\frac{2}{3}}(2\sqrt{K} + P^3)} \tag{3.8}$$

The equilibrium constant and degree of advancement for the steam reforming of methane are presented in Eqs. (3.9) and (3.10). As for ethane, the equilibrium constant is presented in Eq. (3.11) and the degree of advancement for low and high temperatures in Eqs. (3.12) and (3.13). The equilibrium constant and degree of advancement for the steam reforming reaction of natural gas are presented in Eqs. (3.14) and (3.15).

$$K = \frac{3^3\alpha^4P^2}{4 \cdot (1-\alpha^2)^2} \tag{3.9}$$

$$\alpha = \frac{\sqrt{2 \cdot \sqrt{K}}}{\sqrt{3 \cdot \sqrt{3}P + 2\sqrt{K}}} \tag{3.10}$$

$$K = \frac{5^5\alpha^7P^4}{(3+4\alpha)^4(1-\alpha)^3} \tag{3.11}$$

$$\alpha = \sqrt[7]{\frac{3^4K}{5^5P^4}} \tag{3.12}$$

$$\alpha = 1 - \sqrt[3]{\frac{1.3P^4}{K^2}} \tag{3.13}$$

$$K = 0.9K_{\text{steam reforming of methane}} + 0.1K_{\text{steam reforming of ethane}} \tag{3.14}$$

$$\alpha = 0.9_{\text{ steam reforming of methane}} + 0.1\alpha_{\text{steam reforming of ethane}} \tag{3.15}$$

The equilibrium constant and degree of advancement of the steam reforming of methane are presented in Eqs. (3.16) and (3.17), those of the dry reforming of methane in Eqs. (3.18) and (3.19), and those of the steam reforming of biogas in Eq. (3.20) and (3.21).

$$K = \frac{3^3\alpha^4P^2}{4.(1-\alpha^2)^2} \tag{3.16}$$

$$\alpha = \frac{\sqrt{2}\cdot\sqrt{K}}{\sqrt{3\cdot\sqrt{3}P + 2\sqrt{K}}} \tag{3.17}$$

$$K = \frac{4\alpha^4P^2}{(1-\alpha^2)^2} \tag{3.18}$$

$$\alpha = \frac{\sqrt{\sqrt{K}}}{\sqrt{2\cdot P + \sqrt{K}}} \tag{3.19}$$

$$K = 0.669K_{\text{steam reforming}} + 0.331K_{\text{dry reforming}} \tag{3.20}$$

$$\alpha = 0.669_{\text{steam reforming}} + 0.331\alpha_{\text{dry reforming}} \tag{3.21}$$

The equilibrium constant and the degree of advancement of the water-gas *shift* reaction are presented in Eqs. (3.22) and (3.23).

$$K = \frac{\alpha^2}{(1-\alpha)^2} \tag{3.22}$$

$$\alpha = \frac{\sqrt{K}}{1+\sqrt{K}} \tag{3.23}$$

According to Castelan (1986), the degree of advancement (α) increases as the reaction progresses, reaching the limit value when one or more reactants are consumed. This limit value of the degree of advancement refers to the capacity of the reaction mixture to advance.

The behavior analysis of the degrees of advancement at different temperature and pressure ranges of the studied reactions can be observed in Figs. 3.15, 3.16, 3.17, and 3.18.

In the steam reforming of ethanol, it is observed that the degree of advancement is not favored by increased pressure, but by temperature. The best pressure conditions to all studied processes is 1 atm and, as regards temperature conditions, the

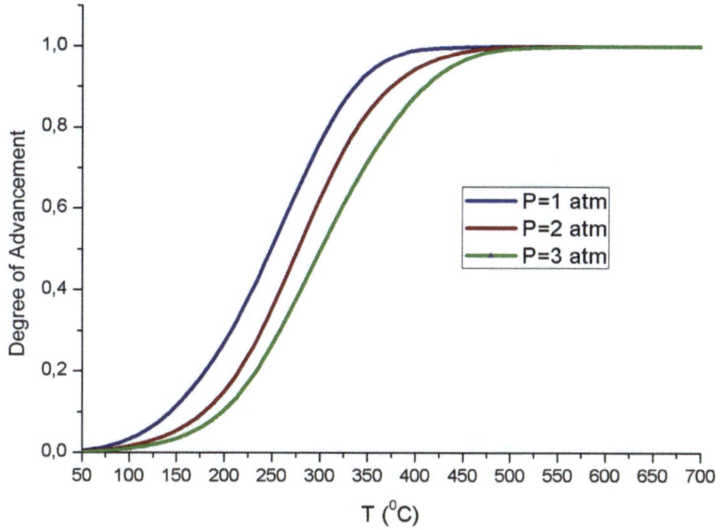

Fig. 3.15 Degree of advancement of the steam reforming of ethanol as a function of temperature under different pressure conditions

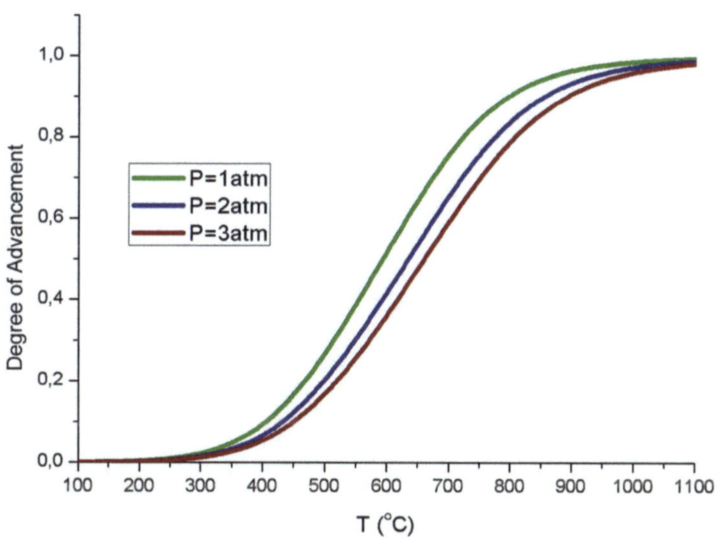

Fig. 3.16 Degree of advancement of the steam reforming of natural gas as a function of temperature under different pressure conditions

steam reforming process of ethanol achieves the highest conversion rate of reagent into products at 400 °C.

The steam reforming of natural gas presented in Fig. 3.15 shows the same behavior as that of the steam reforming of ethanol, but the highest conversion rate occurs at a temperature of 900 °C. The same behavior can be noted in the steam

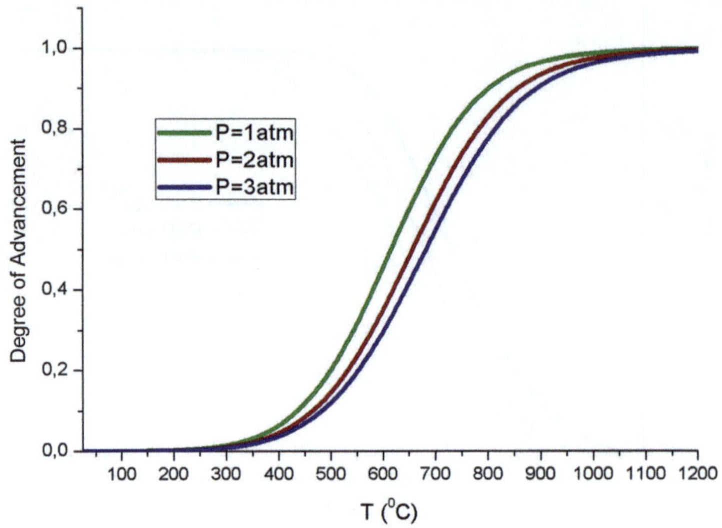

Fig. 3.17 Degree of advancement of the steam reforming of biogas as a function of temperature under different pressure conditions

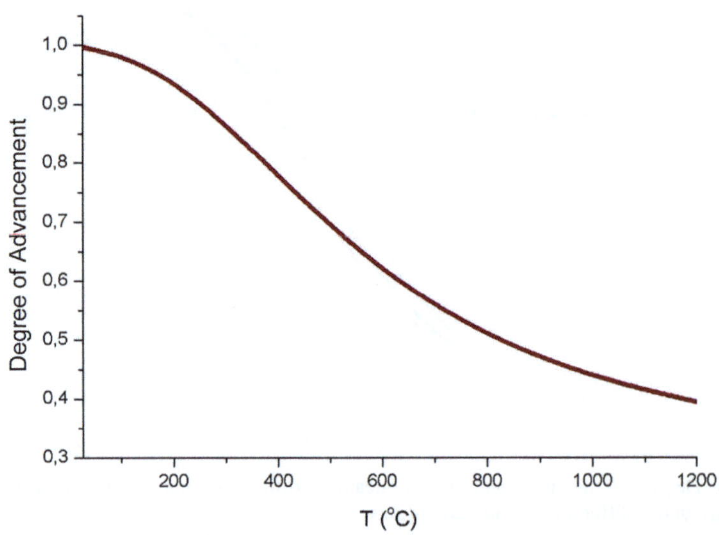

Fig. 3.18 Degree of advancement of the (water-gas) *shift* reforming reaction as a function of temperature

reforming of biogas, as shown in Fig. 3.16, being that the highest conversion rate of this process occurs at 850 °C.

It is observed that, for all steam reforming processes at temperatures above the aforementioned, the conversion rate of reactants into products remains constant.

Figure 3.18 shows that the conversion rate of reactants into products is not favored by temperature increase, and pressure does not influence this process. Thus, at temperatures greater than 200 °C, the conversion rate of reactants into products sharply decreases.

3.1.3 Influence of Catalyst on Studying Steam Reforming Processes

The calculations of the reforming processes' efficiencies were based on the stoichiometry of the reaction. Thus, in order not to consider the reactions with 100 % efficiency and being aware of the necessity of using excess steam (not in stoichiometric proportions of fuel/steam), it was conducted a study of catalysts. The ethanol reforming was based on the catalyst 5 %Ni–5 %Cu/γ-Al_2O_3, the natural gas reforming on 2 %Ru/α-Al_2O_3, and biogas on 5 % Ru/γ-Al_2O_3. In the case of the *shift* reactor, it was based on Cu/ZnO-γ-Al_2O_3. In the study of catalysts and physicochemical analysis, it is noted that the conversion rate indicated by the catalysts in their respective operating temperatures corresponds to the degree of advancement rate calculated at that temperature. For example, the catalyst 5 %Ni-5 %Cu/γ-Al_2O_3 at a temperature of 400 °C has obtained a conversion rate of 97.62 % (MAIA et al. 2007), however, in the physicochemical analysis at the same temperature, there is a degree of advancement of 98 %. Through these studies, it is possible to demonstrate that the catalyst does not change the equilibrium of the reaction, but it only lowers the activation energy required for the reaction to occur. It could be observed the same behavior for the steam reforming processes of natural gas, biogas, and (water-gas) *shift* reaction.

3.1.4 Thermodynamic Efficiency

The thermodynamic efficiency calculated for the steam reforming processes was based on Fig. 3.19.

According to Fig. 3.19, the steam reforming process consists of a steam generator that provides steam to the process and a reforming system which, in this case, consists of a reformer and a *shift* reactor. Hydrogen is the product of the overall process and the fuel used in the steam generator and reforming process is obtained as reagent. Thus, the overall equation of the steam reforming process's efficiency is indicated in Eq. (3.24).

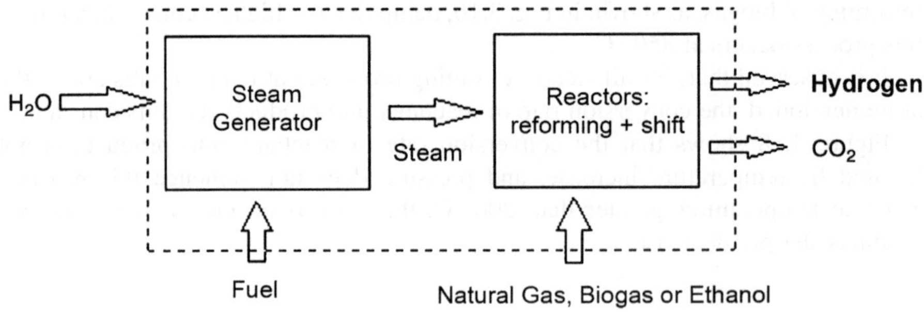

Fig. 3.19 Reforming processes flow chart

$$\eta_{\text{reforming}} = \frac{\dot{m}_{\text{H}_2} \cdot \text{LHV}_{\text{H}_2}}{\dot{m}_{\text{C.C.}} \cdot \text{LHV}_{\text{C.C}} + \dot{m}_{\text{C.R.}} \cdot \text{LHV}_{\text{C.R.}}} \tag{3.24}$$

where

$\eta_{\text{reforming}}$: steam reforming process's efficiency
$\dot{m}$: flow [kg/h]
LHV: lower heating value [kJ/kg]
H_2: hydrogen
C.C.: steam generator fuel
C.R.: reforming fuel

3.1.4.1 Thermodynamic Efficiency of the Steam Reforming Process of Ethanol

In the steam reforming process of ethanol, bagasse was used as fuel to be fed into the steam generator, and ethanol was consumed in the steam reforming reaction.

In Table 3.9, the stoichiometry reactions of the steam reforming process are presented, where both of them occur in reactors 1 and 2 and, subsequently, the overall reaction, with the respective efficiencies of the reactions according to the catalyst chosen for each reactor.

In order to calculate the efficiency of the steam reforming process, the LHV of the reagents and products were utilized, as shown in Table 3.10.

The (volume) flow rate of the reactants and products of the steam reforming process were calculated from the stoichiometry and efficiency of the overall steam reforming equation (Table 3.9) for producing 1 Nm^3/h (0.089 kg/h) of hydrogen, as shown in Table 3.11.

For calculating the flow rate of the bagasse burned in the steam generator, Eq. (3.25) was used:

Table 3.9 Stoichiometric equations of the steam reforming process of ethanol

Type	Stoichiometric reaction	Reaction efficiency (%)
Steam reforming (reactor 1)	$C_2H_6O + H_2O \rightarrow 2CO + 4H_2$	97.62
Shift reactor (reactor 2)	$2CO + 2H_2O \rightarrow 2CO_2 + 2H_2$	93
Overall steam reforming (reactors 1 + 2)	$C_2H_6O + 3H_2O \rightarrow 2CO_2 + 6H_2$	90.78

Table 3.10 Lower heating value of each component of the steam reforming process of ethanol (Adapted from Krona 2012)

Formula	LHV (kJ/kg)
C_2H_6O	28,300
Bagasse	7320
H_2	1,19,950

Table 3.11 (Volume) flow rate values calculated from the ethanol reformer

Formula	$\dot{m}\cdot$ (kg/h)
C_2H_6O	0.3795
H_2	0.0899
CO_2	0.6592

$$\eta_{\text{generator}} = \frac{\dot{m}_{\text{steam}} \cdot \Delta h_{\text{steam}}}{\dot{m}_{\text{bagasse}} \cdot \text{LHV}_{\text{bagasse}}} \tag{3.25}$$

According to Lora and Nascimento (2004), the efficiency of a steam generator being fed with bagasse is 85 %. By considering steam at a temperature of 400 °C and molar ratio of steam/ethanol = 5 (MAIA et al. 2007), the thermodynamic efficiency of the steam reforming process of ethanol is calculated from Eq. (3.26), with obtainment of 73 %.

$$\eta_{\text{reforming}} = \frac{\dot{m}_{H_2} \cdot \text{LHV}_{H_2}}{\dot{m}_{\text{bag.generator}} \cdot \text{LHV}_{\text{bag. generator}} + \dot{m}_{\text{ethane reforming}} \cdot \text{LHV}_{\text{ethane reforming}}} \tag{3.26}$$

3.1.4.2 Efficiency of the Steam Reforming of Natural Gas

In the steam reforming process of natural gas, natural gas was used as fuel, both to be burned in the steam generator and to be consumed by the steam reforming reaction.

For the steam reforming of natural gas (NG), it was considered 90 % methane and 10 % ethane. Table 3.12 presents the stoichiometric reactions that occur in reactors 1 (reformer) and 2 (*shift* reactor), and the overall reaction with the

Table 3.12 Stoichiometric equations of the steam reforming of natural gas

Type	Stoichiometric reaction	Reaction efficiency (%)
Steam reforming of methane + ethane (reactor 1)	$0.9CH_4 + 0.1C_2H_6 + 1.1H_2O \rightarrow 1.1CO + 3.3H_2$	85
Shift reactor (reactor 2)	$1.1CO + 1.1H_2O \rightarrow 1.1CO_2 + 1.1H_2$	93
Overall steam reforming (reactor 1 + reactor 2)	$0.9CH_4 + 0.1C_2H_6 + 2.2H_2O \rightarrow 1.1CO_2 + 4.3H_2$	79.05

Table 3.13 Lower heating value of each component of the steam reforming process of natural gas (Adapted from Krona 2012)

Formula	LHV (kJ/kg)
CH_4	49,934.28
C_2H_6	47,443
Natural gas	49,685.15
H_2	119,950

Table 3.14 (Volume) flow rate values calculated from the natural gas reformer (Adapted from Krona 2012)

Formula	$\dot{m}\cdot$ (kJ/kg)
CH_4	0.1904
C_2H_6	0.0396
Natural gas	0.2300
H_2	0.0899
CO_2	0.5059

respective efficiencies of the reactions, according to the catalyst chosen for each reactor.

For calculating the efficiency of the steam reforming of natural gas, it was used the LHV of reagents and products, as shown in Table 3.13.

The volume flow rate of reactants and products of the steam reforming process was calculated from the stoichiometry and efficiency of the overall steam reforming equation (Table 3.9) for producing 1 Nm^3/h (0.089 kg/h) of hydrogen, which are presented in Table 3.14.

For calculating the flow rate of natural gas used in the steam generator, it was used Eq. (3.27).

$$\eta_{generator} = \frac{\dot{m}_{steam} \cdot \Delta h_{steam}}{\dot{m}_{naturalgas} \cdot LHV_{naturalgas}} \tag{3.27}$$

where

$\eta_{generator}$: steam generator efficiency
$\dot{m}\cdot$: flow [kg/h]
Δh_{steam}: steam enthalpy variation [kJ/kg]

Table 3.15 Stoichiometric equations of the steam reforming of biogas

Type	Stoichiometric reaction	Reaction efficiency (%)
Steam + dry reforming (reactor 1)	$CH_4 + 0.669H_2O + 0.331CO_2 \rightarrow 1.331C + 2.669H_2$	80
Shift reactor (reactor 2)	$1.331CO + 1.331H_2O \rightarrow 1.331CO_2 + 1.331H_2$	93
Overall steam reforming (reactor 1 + 2)	$CH_4 + 2H_2O \rightarrow CO_2 + 4H_2$	74.4

LHV: lower heating value [kJ/kg]

According to Lora and Nascimento (2004), the efficiency of a steam generator being fed with natural gas is 90 %. By considering steam at a temperature of 800 °C and steam/fuel molar ratio = 2.5 (Beurden 2004), the thermodynamic efficiency of the steam reforming process of natural gas is calculated from Eq. (3.28), with obtainment of 71.78 %.

$$\eta_{\text{reforming}} = \frac{\dot{m}_{H_2} \cdot LHV_{H_2}}{\dot{m}_{\text{NGgenerator}} \cdot LHV_{\text{NGgenerator}} + \dot{m}_{\text{NGreforming}} \cdot LHV_{\text{NGreforming}}} \tag{3.28}$$

3.1.5 *Steam Reforming of Biogas*

Biogas was used as fuel in its steam reforming process, both to be burned in the steam generator and to be consumed by the steam reforming reaction.

For the steam reforming of biogas, it was considered two reactions occurring simultaneously on the first reactor, i.e., the steam and dry reforming reactions. Table 3.15 presents the stoichiometric reactions that occur in reactors 1 (reformer) and 2 (shift reactor) and, subsequently, the overall reaction, with the respective efficiencies of the reactions, according to the catalyst chosen for each reactor.

For calculating the efficiency of the steam reforming calculations, it was used the LHV of reagents and products, as shown in Table 3.16.

The volume flow rate of reactants and products of the steam reforming process were calculated from the stoichiometry and efficiency of the overall steam reforming equation (Table 3.9) for producing 1 Nm^3/h (0.089 kg/h) of hydrogen, which are presented in Table 3.17.

For calculating the volume flow rate of biogas used in the steam generator, it was used Eq. 3.29:

Table 3.16 Lower heating value of each component in the steam reforming process of biogas (Adapted from Krona 2012)

Formula	LHV (kJ/kg)
CH_4	49,934.28
CO_2	0
Biogas	33,406
H_2	119,950

Table 3.17 (Volume) flow rate values calculated from the biogas reformer

Formula	$\dot{m}\cdot$ (kg/h)
CH_4	0.2416
$CO_{2\ (reactant)}$	0.2417
Biogas	0.4723
H_2	0.0899
$CO_{2(product)}$	0.4945

$$\eta_{\text{generator}} = \frac{\dot{m}_{\text{steam}} \cdot \Delta h_{\text{steam}}}{\dot{m}_{\text{biogas}} \cdot \text{LHV}_{\text{biogas}}} \tag{3.29}$$

According to Lora and Nascimento (2004), the efficiency of a steam generator being fed with biogas is 90 %. Considering steam at a temperature of 700 °C and Gibbs free energy = 3.04 (AVRAAM et al. 2010), the thermodynamic efficiency of the steam reforming process of biogas is calculated from Eq. (3.30), with obtainment of 68 %.

$$\eta_{\text{reforming}} = \frac{\dot{m}_{H_2} \cdot \text{LHV}_{H_2}}{\dot{m}_{\text{biog. generator}} \cdot \text{LHV}_{\text{biog. generator}} + \dot{m}_{\text{biog. generator}} \cdot \text{LHV}_{\text{biog. reforming}}} \tag{3.30}$$

3.1.6 *Plasma Reforming*

The most commonly known H_2 production processes through a fuel reforming reaction are steam, partial oxidation, and autothermal reforming. However, these reforming processes involve various technical limitations, such as slow operation characteristics, requirement of large external thermal sources due to strong thermal absorption reactions, and poisonous catalysts containing a small amount of sulfur in the fuel.

Therefore, to overcome such limitations, a plasma reforming method that uses an electric discharge has been receiving attention. It uses electrical energy and is effective at transforming various fuels because there is no generation of additional polluting impurities. In particular, electric discharge typically creates electrons with high energy, can easily transform compound materials with lower reaction rates, and, because the plasma reaction takes place in any volume capacity held by plasma electric discharge, the reactor is compact and effective.

Plasma contains reactive radicals, ions, and high-energetic electrons. The high reactivity shown by these species enhances the chemical reaction rates, whereby expensive and impure vulnerable catalysts can be avoided. This plasma capability, as well as its high energy density, ensures the compactness of the plasma system. Furthermore, it can be adapted so as to reform various hydrocarbons, such as natural gas, gasoline, heavy oils, and biofuels. Another advantage is the quick response time that can be achieved on account of the process being powered by electricity. When steam is used as the plasma's supporting gas, reductive and oxidative radicals such as H, OH, and O are produced in it, thus enabling it to be effective at reforming different hydrocarbons.

Recently developed microwave plasma sources (MPSs), when operated at atmospheric pressure, seem to have an enormous potential for hydrogen production via hydrocarbon reforming. Microwave plasma reforming at atmospheric pressure is one of the techniques providing electron temperatures of 4000–10,000 K and heavy particle temperatures of 2000–6000 K. Such properties of microwave plasma provide proper conditions for methane reforming.

Among the alternative reforming techniques, plasma-assisted fuel reforming appears to be particularly attractive for automotive applications, especially regarding response time, compactness, and absence of a catalyst. Previous works carried out at MIT based on thermal plasma have demonstrated high efficiency in term of conversion rates. Unfortunately, these systems have shown very poor energy conversion efficiency, as a result of the high energy consumption of thermal plasmas.

3.2 Electrolytic Processes

For calculating the efficiency of theoretical electrolytic processes, it was based on the electrolyzer efficiency and on the average efficiencies of wind, photovoltaic, and hydroelectric power plants. The electrolyzer efficiency was calculated through a commercially available electrolyzer. It was considered that the electricity used to power this electrolyzer is the one that would be produced by plants in periods of low consumption, as shown in Fig. 3.20.

The thermodynamic efficiencies for electrolysis using the electricity generated by wind, photovoltaic, and hydroelectric power plants were calculated according to Eqs. (3.31)–(3.33), respectively.

$$\eta_{\text{electro.plant.wind.}} = \eta_{\text{electrolyzer}} \cdot \eta_{\text{windfarm}} \tag{3.31}$$

$$\eta_{\text{electro.plant.photovoltaic.}} = \eta_{\text{electrolyzer}} \cdot \eta_{\text{photovoltaic plant}} \tag{3.32}$$

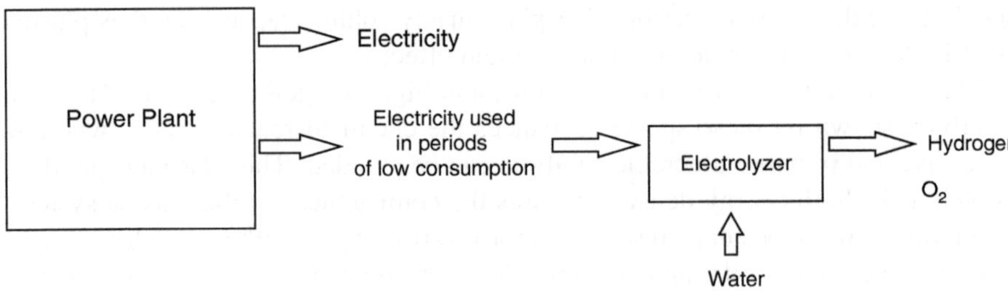

Fig. 3.20 Electrolytic process flow chart

$$\eta_{\text{electro.plant.hydroelectric.}} = \eta_{\text{electrolyzer}} \cdot \eta_{\text{hydroelectric plant}} \tag{3.33}$$

3.2.1 *Electrolyzer*

For calculating the thermodynamic efficiency of the electrolyzer, it was selected a model that relies on national expertise and assistance. The equipment was designed by the Italian company Nitidor, but it is marketed by Hytron, as shown in Fig. 3.21. The basic technical features of the selected model can be seen in Table 3.18.

As regards the Nitidor electrolyzer, its thermodynamic efficiency is calculated using Eq. (3.34), where 84.4 % is obtained.

$$\eta_{\text{electrolizer}} = \frac{\dot{m}_{H_2} \cdot PCI_{H_2}}{Econs} \tag{3.34}$$

where

$\eta_{\text{electrolizer}}$: electrolyzer thermodynamic efficiency
Econs: electrolyzer power consumption

3.2.2 *Thermodynamic Efficiency of Electrolysis Using Wind Power*

According to Gipe (1995) and Terciote (2002), the efficiency of commercial wind power plants is around 35 % for large-scale systems.

Thus, by considering around 35 % efficiency of wind power plants, it is found that the efficiency of hydrogen production process using energy from this type of system is 29.5 %.

Fig. 3.21 Control system (on the *left*) and electrolyzer (on the *right*) (Reproduced from the Nitidor product catalog 2011)

Table 3.18 Selected features of the electrolyzer (Reproduced from the Nitidor product catalog 2011)

Hydrogen production capacity	1 Nm^3/h
Nominal capacity	1.1 Nm^3/h
Theoretical electricity consumption	3.55 kW
Distilled water consumption (<5 μS/cm)	1 L/h

3.2.3 Electrolysis Using Solar Energy

The average efficiency of photovoltaic panels or power generation plants using such a device is low, i.e., around 14 % (ANEEL 2008).

Thus, by considering around 14 % efficiency of photovoltaic power plants, it is found that the efficiency of hydrogen production process using energy from this type of system is 11.8 %.

3.2.4 Electrolysis Using Hydroelectric Power

According to Sampaio et al. (2005) and Siqueira (2006), the efficiency of electricity generation in hydropower plants is around 95 %.

Thus, by considering 95 % average efficiency of a hydroelectric power plant, it is found that the efficiency of hydrogen production process with electricity using this type of system is 80 %.

3.3 Hydrogen Production from Algae

According to the methodology used to produce 1 Nm^3/h of hydrogen, the alga *C. reinhardtii* and direct biophotolysis was considered. The flow chart containing the hydrogen production from algae is shown in Fig. 3.22.

According to Fig. 3.22, hydrogen production is based on two types of bioreactors, one for algae growth and five others for hydrogen production. In both cases, closed reactors were considered in order to avoid contamination. The algae growth reactor is fed with the TAP medium through a pump, and air is pumped with a compressor. Now, hydrogen production reactors are fed with 90 % TAP-S and 10 % of the medium with algae. At the end of each period of hydrogen production, five reactors are emptied and their content is directed to a storage tank through gravity. All reactors use sunlight, therefore, they are exposed to light for 12 h, and for 12 more hours without it afterwards. According to literature, when there is a change in periods of light and darkness, both efficiencies of algae growth and hydrogen production are not compromised (Kosourov et al. 2002).

The method for transferring the medium contained in the algae reactor to the reactors for hydrogen production can be either centrifugation or dilution. In the presently analyzed case, it was opted for the dilution method with no need for a pump (i.e., through gravity), aiming at the sustainability of the process.

The maximum theoretical rate of hydrogen production is 20 g m^{-2} day^{-1} by using sunlight (Bhutto et al. 2011; Melis and Happe 2001), being that, only 45 % of the solar spectrum is located in the region of visible radiation which is used in the photosynthesis process (Melis 2008; Berberoglu and Lilon 2010). The solar spectrum of Brazil, in most regions, presents an annual radiation average of kWh m^{-2} day^{-1} (Solarpaces 2013), whereas by considering 45 % of radiation, 2.47 kWh m^{-2} day^{-1} of it is absorbed by algae, both for their growth and their period of adaptation to an anaerobic medium and for hydrogen production. The maximum theoretical rate of hydrogen production is 20 g m^{-2} day^{-1} by using sunlight (Bhutto et al. 2011; Melis and Happe 2001), being that, only 45 % of the solar spectrum is located in the region of visible radiation which is used in the photosynthesis process (Melis 2008; Berberoglu and Lilon 2010). The solar spectrum of Brazil, in most regions, presents an annual radiation average of 5.5 kWh m^{-2} day^{-1} (Solarpaces 2013), whereas by

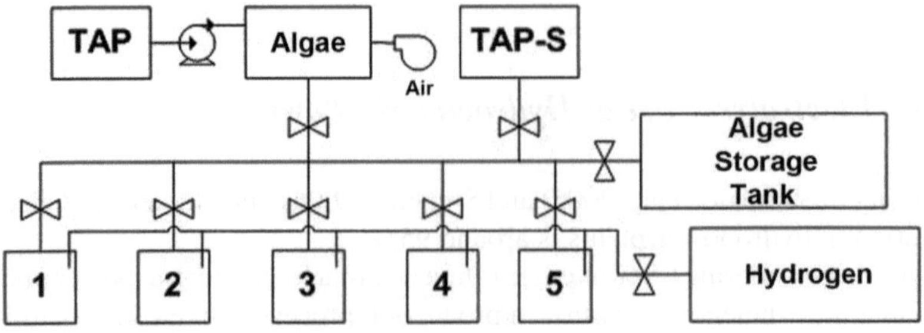

Fig. 3.22 Hydrogen production from microalgae flow chart

considering 45 % of radiation, 2.47 kWh m^{-2} day^{-1} of it is absorbed by algae, once again both for their growth and their period of adaptation to an anaerobic medium and for hydrogen production.

There are several types of reactors in literature as aforementioned. Closed reactors were considered for both the phases of algae growth and hydrogen production. Such reactors feature optimal efficiencies due to their high surface-to-volume ratio (Burgess et al. 2011). For all reactors, it was considered a depth of 5 cm.

3.3.1 Thermodynamic Efficiency of Hydrogen Production from Algae

The efficiency of hydrogen production process from the green alga *C. reinhardtii* was calculated by Eq. (3.35).

$$\eta_{\text{system}} = \frac{E_{\text{hydrogen}}}{e_{\text{growth}} + e_{\text{adapt.}} + e_{\text{prod.hydro}}} \tag{3.35}$$

where:

E_{hydrogen}: energy stored from hydrogen production (140 h) [kJ]
e_{growth}: energy required for the period of algae growth (60 h) [kJ]
$e_{\text{adapt.}}$: energy required for the period of algae adaptation to an anaerobic medium (24 h) [kJ]
$e_{\text{prod. hydro}}$: required energy supply in the hydrogen production period (140 h) [kJ]

The following data was considered for this analysis:

$\dot{m}_{\text{hydrogen}}$: 1 Nm^3/h (hydrogen production rate)
LHV_{hydrogen}: 119,950 [kJ/kg] (Krona 2012).

A period of 60 h (2–3 days) is required for algae growth (Stephanie et al. 2011).

As regards the period of hydrogen production, according to literature, the alga *C. reinhardtii* can withstand the condition of anaerobiosis for 5–6 days (140 h) (Ahmed et al. 2011; Ghirardi et al. 2001; Kosourov et al. 2002). Then, it was considered a period of 140 h for hydrogen production.

For calculating the absorbed solar energy, according to period and area of exposure, it was adopted a hydrogen production maximum rate of 20 g m^{-2} day^{-1} , as the one available in literature (Bhutto et al. 2011; Melis and Happe 2001). Thus, the area calculated for the period of production of 1 Nm^3/h of hydrogen was 107.88 m^2 (five reactors being 4.31 m wide and 5 m long). Given that 10 % of the volume of hydrogen production reactors are from the algal growth reactor, it is concluded that the area required to ensure the production of 1 Nm^3/h of hydrogen is

10.79 m^2. Therefore, the values of radiation absorbed by period and area were: e_{growth}: $2.4 \cdot 10^5$ kJ, $e_{adapt.}$: $9.6 \cdot 10^5$ kJ, and $e_{prod.\,hydro}$: $5.7 \cdot 10^6$ kJ.

As regards the energy stored from hydrogen production, it was considered Eq. (3.36).

$$E_{Hydrogen} = \dot{m}_{Hydrogen} \cdot \mathrm{LHV} \cdot \mathrm{HH} \quad (3.36)$$

Given that the adopted hydrogen production period (HH) was 140 h, it is concluded that $E_{Hydrogen}$: $1.5 \cdot 10^6$ kJ.

Thus, the total efficiency found for this type of production process is 22 % which is close to the theoretical values reported in literature (Esper et al. 2006).

3.4 Energy Efficiencies Comparison

The energy efficiency values of the production process described above can be observed in Fig. 3.23.

According to Fig. 3.23, the highest efficiency value among the hydrogen production processes are those of the steam reforming systems (ethanol, natural gas, and biogas) and the electrolysis from hydroelectric power plants. Electrolytic systems (photovoltaic and wind power) are affected by the low conversion efficiencies of energy into electricity. Among the stem reforming processes, the one which has obtained the highest efficiency value is the steam reforming of ethanol. One of the reasons for the steam reforming of ethanol being the most energy-

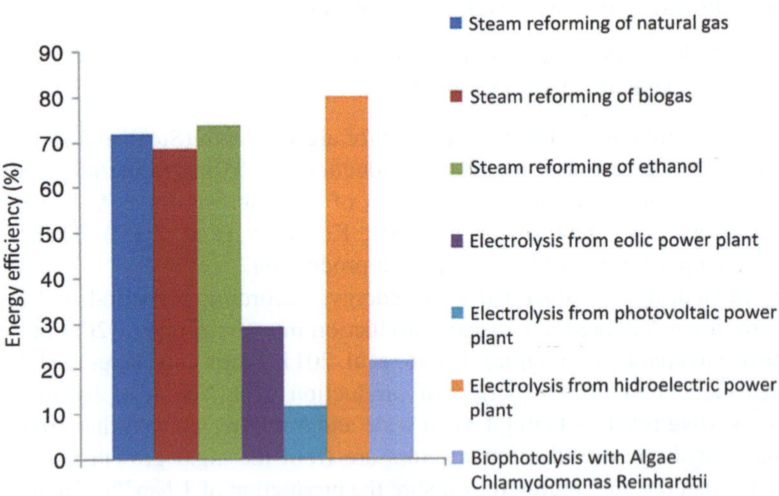

Fig. 3.23 Energy efficiency comparison of hydrogen production processes

efficient process is attributed to its lower operating temperatures (400 °C) when compared to the steam reforming processes of natural gas (900 °C) and biogas (850 °C).

Despite the fact that the steam reforming of natural gas is operated at higher temperature conditions than the steam reforming of biogas is, when comparing these two processes from an energetic point of view, it is found that the former is more efficient. This fact is justified by the efficiency of the natural gas steam reforming reaction (79.05 %) being greater than that of the biogas process (74.4 %), which is influenced by the amount of methane in the composition of natural gas (90 %) and biogas (75.7 %).

References

Adamin R (2002) Almendra E Físico-quimca: Uma aplicação aos materiais. Coppe/UFRJ, Rio de Janeiro. ISBN: 85-285-0057-8

Ahmed A, Abdel M, Farag MAS (2011) Roles of microalgae and bacteria in hydrogen production as one of the renewable energy resources. In: Bulletin for environmental researches, vol 12, no 2. Assiut University

Avraam DG, Halkides TI, Liguras DK, Bereketidou OA, Goula MA (2010) An experimental and theoretical approach for the biogas steam reforming reaction. Int J Hydrogen 35:9818–9827

ANEEL Brasil (2008) Atlas de Energia Eletrica do Brasil: Agencia Naciona de Energia Eletrica. ANEEL, Brasilia, p 236

Berberoglu H, Lilon L (2010) Maximizing the solar to H2 conversion efficiency of outdoor photobioreactors using mixed cultures. Int J Hydrogen Energy 35:500–510

Beurden PV (2004) On the catalytic aspects of steam-methane reforming: a literature survey. ECN-I-04-003

Bhutto AW, Bazmi AA, Kardar MN, Yaseen M (2011) Developments in hydrogen production through microbial processes; Pakistan's prospective. Int J Chem Environ Eng 2(3):189–205

Burgess SJ, Tamburic B, Zemichael F, Hellgardt K, Nixon PJ (2011) Solar driven hydrogen production in green algae. Adv Appl Microbiol 75:71–110

Castelan G (1986) Fundamentos da Físico/Química. Editora LTC, Rio de Janeiro, 529p

Esper B, Badura A, Rogner M (2006) Photosynthesis as a power supply for bio- hydrogen production. Trends Plant Sci 11:543–549

Ghirardi ML, Kosourov S, Seibert M (2001) Cyclic photobiological algal h2-production. In: Proceedings of the 2001 DOE Hydrogen Program Review, NREL/CP-570-30535

Gipe P (1995) Wind energy—comes on age. Wiley, New York, 536 p. http://books.google.com.br/books/about/Wind_Energy_Comes_of_Age.html?id=8itBNxBL4igC&redir_esc=y. Accessed 11 Jan 2014

Kosourov S, Tsygankov A, Seibert M, Ghirardi ML (2002) Sustained hydrogen photoproduction by *Chlamydomonas Reinhardtii*: effects of culture parameters. Biotechnol Bioeng 78:731–740

Krona. http://www.krona.srv.br/display05.htm Accessed 10 Oct 2012

Lora EES, Nascimento MAR (2004) Geração Termelétrica: Planejamento, Projeto e Operação, vol 2, Rio de Janeiro. In: Interciência

Maia TA, Bellido JDA, Assaf EM (2007) Produção de hidrogênio a partir da reforma a vapor de etanol utilizando catalisadores Cu/Ni/ γ-Al_2O_3. Quim Nova 30:339–345

Melis A (2008) Solar energy conversion efficiencies in photosynthesis: minimizing the chlorophyll antennae to maximize efficiency. Plant Sci 177:272–280

Melis A, Happe T (2001) Hydrogen production. Green algae as a source of energy. Plant Physiol 127:740–748

Sampaio LMB, Ramos FS, Sampaio Y (2005) Privatização e eficiência das usinas hidrelétricas brasileiras. Econ Aplic 9:465–480

Siqueira RBP (2006) Construção de diagramas de custos para PCH incorporando turbinas de mercado. 151 f. Dissertação (Mestrado)—Curso de Engenharia Mecânica, Departamento de Energia, Unesp, Guaratinguetá

Solarpaces. http://www.solarpaces.org/News/Projects/Brazil.htm. Accessed 20 Mar 2013

Sosa MI (2012) Análisis Termoeconómico del proceso de reforma a vapor de gás natural. Tese (Doutorado). Universidad Nacional de La Plata, UNLP

Stephanie CG, Sabine H, Konstantin P, Moritz H, Christian W, Rainer B (2011) Outdoor cultivation of Chlamydomonas Reinhardtii for photobiological hydrogen production. J Appl Phycol 24:319–327

Terciote R (2002) Eficiência energética de um sistema eólico isolado. An.4. Enciclopedia. Energética. Meio Rural

Chapter 4
Economic Studies of Some Hydrogen Production Processes

Lúcia Bollini Braga, Celso Eduardo Tuna, Fernando Henrique Mayworm de Araujo, Lucas Fachini Vane, Daniel Travieso Pedroso, and José Luz Silveira

4.1 Hydrogen Production Cost

The cost of hydrogen production Eq. (4.1) incurs expenditures on investment (steam generator, reformer, electrolyzer, and reactors), operation, and maintenance. In all processes, investment in a few items of equipment such as pumps and valves, among others, was disregarded.

$$C_{H_2} = \frac{Invf}{HE_{H_2}} + C_{op} + C_{man} \tag{4.1}$$

where:

C_{H_2}—Hydrogen production cost [US$/kWh]
Inv—investment costs [US$]
H—equivalent utilization period [h/year]
f—annuity factor [1/year]
C_{op}—operation cost [US$/kWh]
C_{man}—maintenance cost [US$/kWh]

L.B. Braga • C.E. Tuna • F.H.M. de Araujo • L.F. Vane • D.T. Pedroso
Group of Optimization of Energy Systems—GOSE, College of Engineering of Guaratinguetá, Institute of Bioenergy Research—IPBEN, São Paulo State University—UNESP, Dr. Ariberto Pereira da Cunha Ave, 333, Guaratinguetá 12516-410, SP, Brazil

J.L. Silveira (✉)
Group of Optimization of Energy Systems—GOSE, College of Engineering of Guaratinguetá, Institute of Bioenergy Research—IPBEN, São Paulo State University—UNESP, Dr. Ariberto Pereira da Cunha Ave, 333. Pedregulho, Guaratinguetá 12516-410, São Paulo, Brazil
e-mail: joseluz@feg.unesp.br

J.L. Silveira (ed.), *Sustainable Hydrogen Production Processes*, Green Energy and Technology, DOI 10.1007/978-3-319-41616-8_4

The energy available from the hydrogen production processes is calculated by Eq. (4.2).

$$E_{H_2} = \dot{m}_{H_2} \cdot PCI_{H_2} \tag{4.2}$$

4.1.1 Maintenance and Operation Costs

The operation cost depends on the energy source and/or fuel used in the process. In the case of the reforming processes, it is considered the fuel consumed by the steam generator and the fuel used in the reforming process, as Eq. (4.3). In the case of electrolysis processes, it is considered the cost of electricity consumption (wind, photovoltaic, or hydroelectric power), as shown in Eq. (4.4). As for the operation cost of the biological process, it is considered the media expenditures, as Eq. (4.5).

$$C_{op} = \frac{E_{fuel.generator} \cdot P_{fuel.generator}}{E_{H_2}} + \frac{E_{fuel.reform.} \cdot P_{fuel.reform.}}{E_{H_2}} \tag{4.3}$$

$$C_{op} = \frac{E_{Elect_{(cons)}} \cdot P_{Elect}}{E_{H_2}} + \frac{\dot{m}_{H_2O} \cdot C_{H_2O}}{E_{H_2}} \tag{4.4}$$

$$C_{op} = \frac{(growthmedium) + (productionmedium)}{E_{H_{2year}}} \tag{4.5}$$

where:
C_{H_2O}—water cost [US\$/m^3]
$Cgrowthmedium$—cost of growth medium added at every 164 h [US\$/year]
$Cproductmedium$—cost of replenishing the algae production medium at every 164 h [US\$/year]
E_{fuel}—energy supplied by the fuel [kW]
$E_{Elect_{(cons)}}$—required energy supply for the electrolyzer [kW]
$E_{H_{2year}}$—energy stored from hydrogen production [kW/year]
$\dot{m}_{H_2O}$—required water flow for the electrolyzer [m^3]
P_{Fuel}—fuel costs [US\$/kWh]
P_{Elect}—electricity cost [US\$/kWh]

According to Kothari et al. (2008), the maintenance cost of the reforming system can be considered as being 3 % of the investment cost, as shown in Eq. (4.6).

$$C_{man} = 0.03 \cdot \frac{Invf}{HE_{H_2}} \tag{4.6}$$

4.1.2 Annuity Factor

The annuity factor is calculated according to Eqs. (4.7) and (4.8).

$$f = \frac{q^k(q-1)}{q^k - 1} \tag{4.7}$$

$$q = 1 + \frac{r}{100} \tag{4.8}$$

where:

f—annuity factor [1/year]
k—year
r—interest rate [%]

4.2 Economic Analysis of Process

The items below were taken into consideration in this study:

- Investment in the reforming system (according to the prototypes developed at LOSE—Laboratory for Optimization of Energy Systems): $20,000.00 USD for the ethanol reformer (P&D CEMIG) and $29,488.26 USD for the biogas reformer (Fapesp PPPP-process 06/52260-2). For the natural gas reformer, it was adopted the same cost of the biogas reformer, due to the similarity between the processes.
- Annual interest rate: 4, 8 and 12 % (BCB 2013)
- Periods of operation: 2.920, 3.650, and 4.380 h/year of operation (being 8, 10, and 12 h a day for 365 days). These periods of operation were established to enable a comparison between all the studied processes, whereas the maximum sun exposure in Brazil is 12:00 (Webcalendar 2014)
- The LHV of the used fuels and H_2 in [kJ/kg]
- Methane: 49.934, 28 (Krona 2012)
- Ethane: 47.443 (Krona 2012)
- Ethanol: 28.300 (Silva 2010)
- Bagasse: 7.320 (Silva 2010)
- H_2: 119.950 kJ/kg (Krona 2012)

According to Braga et al. (2013), the cost of generating biogas is $0.0518 USD/kWh. The cost of natural gas is $0.9 USD/Nm^3 ($0.08 USD/kWh), according to COMGAS (2013), and the cost of ethanol and bagasse is $0.22 USD/l ($0.3471 USD/kWh) and $7.17 USD/ton ($0.0035 USD/kWh), respectively, according to Silva (2010).

For the calculations, it was estimated a yield of 90 % for the steam generator (Lora and Nascimento 2004) and 85 % for the steam generator using bagasse (Lora and Nascimento 2004). The power consumption of the reforming system was calculated according to Eq. (4.9).

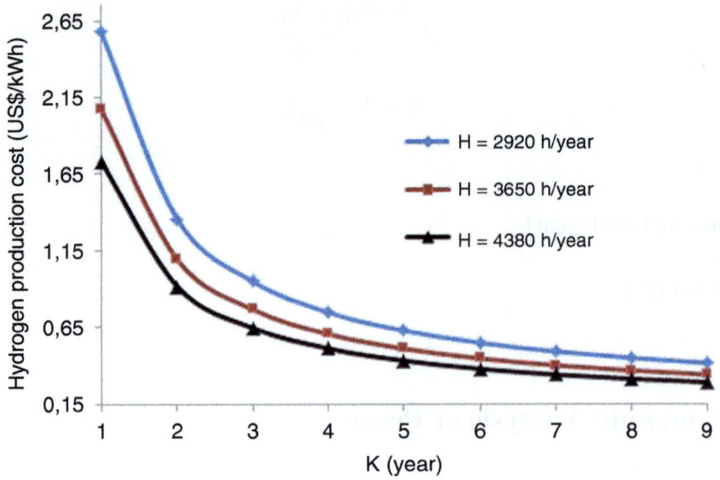

Fig. 4.1 Hydrogen production cost of the steam reforming of ethanol as a function of payback ($r = 8\,\%$)

$$\eta_{\text{generator}} = \frac{E_{\text{steam}}}{E_{\text{fuel}}} \tag{4.9}$$

where:

$\eta_{\text{generator}}$—steam generator yield [%]
E_{fuel}—energy supplied by the fuel in the steam generator [kW]
E_{steam}—steam power [kW]

Analyzing the production of 1 Nm^3/h of hydrogen, it was used the fuel flow rates previously calculated in the efficiency chapter, both for the reformer and the steam generator.

Given the aforementioned, the following results were obtained:

The charts containing the production costs of producing 1 Nm^3/h of hydrogen from the ethanol reforming are in Figs. 4.1 and 4.2.

According to Fig. 4.1, by setting the interest rate at 8 %, it is observed that, for this type of process, the hydrogen production costs for an operation period of 2.920 h/year started at $2.58 USD/kWh ($k = 1$) and, on the 9th year, it was in around $0.41 USD/kWh. For an operation period of 3.650 h/year, it started at $2.07 USD/kWh ($k = 1$) and, on the 9th year, it was in around $0.34 USD/kWh. For an operation period of 4.380 h/year, it started at $1.73 USD/kWh ($k = 1$) and, on the 9th year, it was in around $0.29 USD/kWh.

According to Fig. 4.1, by setting the interest rate at 8 %, it is observed that, for this type of process, the hydrogen production costs for an operation period of 2.920 h/year started at $2.58 USD/kWh ($k = 1$) and, on the 9th year, it was around $0.41 USD/kWh. For an operation period of 3.650 h/year, it started at $2.07 USD/kWh ($k = 1$) and, on the 9th year, it was around $0.34 USD/kWh. For an operation

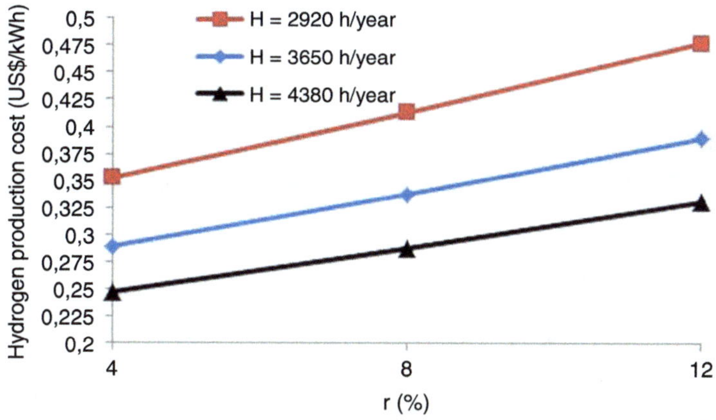

Fig. 4.2 Hydrogen production cost of the steam reforming of ethanol as a function of annual interest rate ($k = 9$ years)

period of 4.380 h/year, it started at \$1.73 USD/kWh ($k = 1$) and, on the 9th year, it was around \$0.29 USD/kWh.

According to Fig. 4.2, by varying the annual interest rate in 4, 8, and 12 % and keeping an operation period of 9 years, the hydrogen production costs for this type of process were:

- \$0.35; 0.29; and 0.25 USD/kWh at an annual interest rate of 4 % and operation periods of 2.920, 3.650, and 4.380 h/year, respectively
- \$0.41; 0.34; and 0.29 USD/kWh at an annual interest rate of 8 % and operation periods of 2.920, 3.650, and 4.380 h/year, respectively
- \$0.48; 0.39; and 0.33 USD/kWh at an annual interest rate of 12 % and operation periods of 2.920, 3.650, and 4.380 h/year, respectively

The charts containing the production costs of 1 Nm^3/h of hydrogen from the natural gas reforming are in Figs. 4.3 and 4.4.

According to Fig. 4.3, by setting the annual interest rate at 8 %, it is observed that, for this type of process, the hydrogen production costs for an operation period of 2.920 h/year started at \$3.87 USD/kWh ($k = 1$) and, on the 9th year, it was around \$0.67 USD/kWh. For an operation period of 3.650 h/year, it started at \$3.12 USD/kWh ($k = 1$) and, on the 9th year, it was around \$0.56 USD/kWh. For an operation period of 4.380 h/year, it started at \$2.62 USD/kWh ($k = 1$) and, on the 9th year, it was around \$0.49 USD/kWh.

According to Fig. 4.4, by varying the annual interest rate in 4, 8, and 12 % and keeping an operation period of 9 years, the hydrogen production costs for this type of process were:

- \$0.58; 0.49; and 0.43 USD/kWh at annual interest rate of 4 % and operation periods of 2.920, 3.650, and 4.380 h/year, respectively

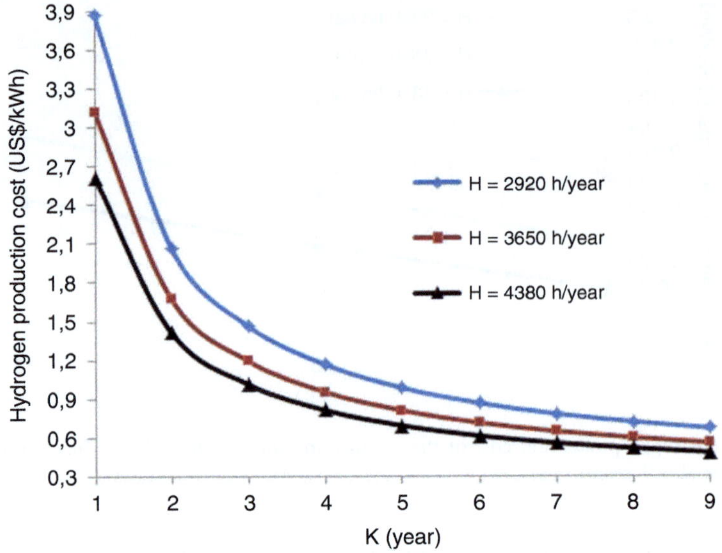

Fig. 4.3 Hydrogen production cost of the steam reforming of natural gas as a function of *payback* ($r = 8\,\%$)

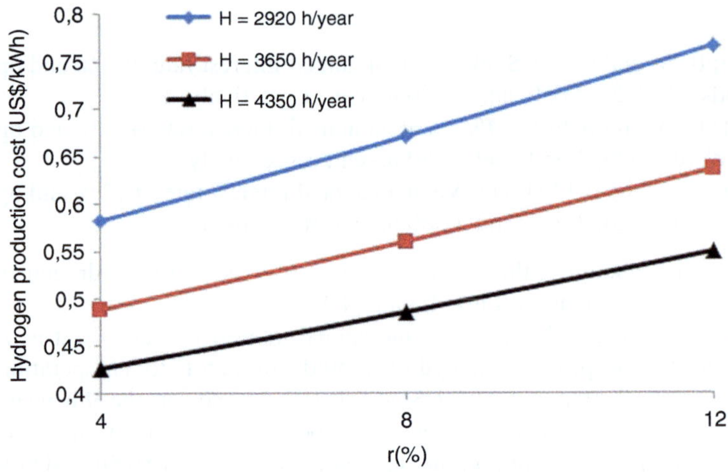

Fig. 4.4 Hydrogen production cost of the steam reforming of natural gas as a function of annual interest rate ($k = 9$ years)

- $0.67; 0.56; and 0.49 USD/kWh at an annual interest rate of 8 % and operation periods of 2.920, 3.650, and 4.380 h/year, respectively
- $0.77; 0.64; and 0.55 USD/kWh at an annual interest rate of 12 % and operation periods of 2.920, 3.650, and 4.380 h/year, respectively

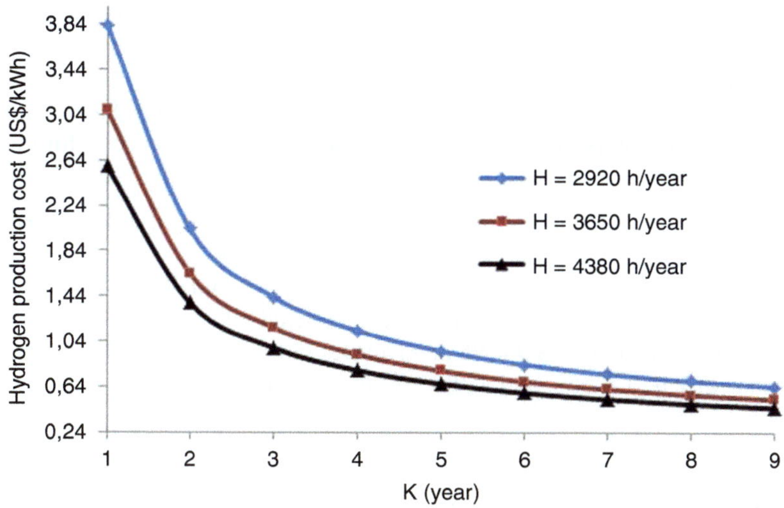

Fig. 4.5 Hydrogen production cost of the steam reforming of biogas as a function of *payback* ($r = 8\,\%$)

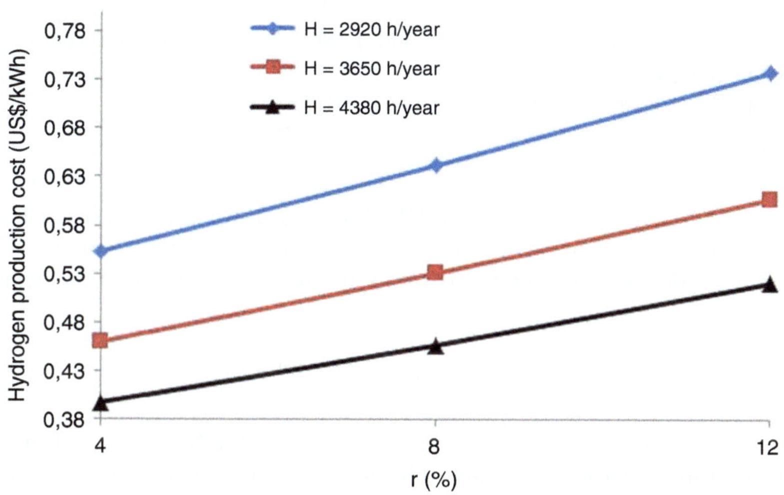

Fig. 4.6 Hydrogen production cost of the steam reforming of biogas as a function of annual interest rate ($k = 9$ years)

The charts containing the production costs of 1 Nm^3/h of hydrogen from the biogas reforming are in Figs. 4.5 and 4.6.

According to Fig. 4.5, by setting the annual interest rate at 8 %, it is observed that, for this type of process, the hydrogen production costs for an operation period of 2.920 h/year started at $3.84 USD/kWh ($k = 1$) and, on the 9th year, it was around $0.64 USD/kWh. For an operation period of 3.650 h/year, it started at $3.09/

kWh ($k = 1$) and, on the 9th year, it was around \$0.53 USD/kWh. For an operation period of 4.380 h/year, it started at \$2.59 USD/kWh ($k = 1$) and, on the 9th year, it was around \$0.46 USD/kWh.

According to Fig. 4.6, by varying the annual interest rate in 4, 8, and 12 % and keeping an operation period of 9 years, the hydrogen production costs for this type of process were:

- \$0.55; 0.46; and 0.40 USD/kWh at an annual interest rate of 4 % and operation periods of 2.920, 3.650, and 4.380 h/year, respectively
- \$0.64; 0.53; and 0.46 USD/kWh at an annual interest rate of 8 % and operation periods of 2.920, 3.650, and 4.380 h/year, respectively
- \$0.74; 0.61; and 0.52 USD/kWh at an annual interest rate of 12 % and operation periods of 2.920, 3.650, and 4.380 h/year, respectively

4.3 Economic Analysis of Electrolytic Process

The items below were taken into consideration in this study:

- Annual interest rate: 4, 8 and 12 % (BCB 2013)
- Periods of operation: 2.920, 3.650, and 4.380 h/year of operation (being 8, 10, and 12 h a day for 365 days)
- The LHV of H_2 at 119.950 kJ/kg (Krona 2012)
- Required energy supply for the electrolyzer: 5.5 kW (Nitidor 2013)

The investment in the electrolyzer was based on Eq. (4.10) which was developed by Saur (2008) and that, according to NREL (National Renewable Energy Laboratory), the investment of an electrolyzer is estimated to be at \$1000 USD for hydrogen production ranging from 0.1 to 100 kg/h with an amazing degree of accuracy.

$$C_{\text{elect}} = 224.49 \cdot \left(\dot{m}_{\text{H}_2}\right)^{0.6156} \tag{4.10}$$

As the present study is based on the production of 1 Nm^3/h of hydrogen and considering the density as being 0.0899 kg/m^3, the investment cost of the electrolyzer was considered to be at \$5633.91 USD.

It was considered that the plants are already in operation, as well as all necessary items of equipment, thus the operation cost is related to the price of electricity from wind, photovoltaic, and hydroelectric power plants, and to the cost of water needed for feeding the electrolyzer.

The water supply needed to feed the electrolyzer was obtained stoichiometrically by considering the efficiency of the system, thence it was obtained $\dot{m}_{\text{H}_2\text{O}} = 0.9\,\text{L/h}$. In order to draw comparisons among the wind, solar, and hydroelectric sources, minimum and maximum costs of such power supplies were adopted, based on Fig. 4.7.

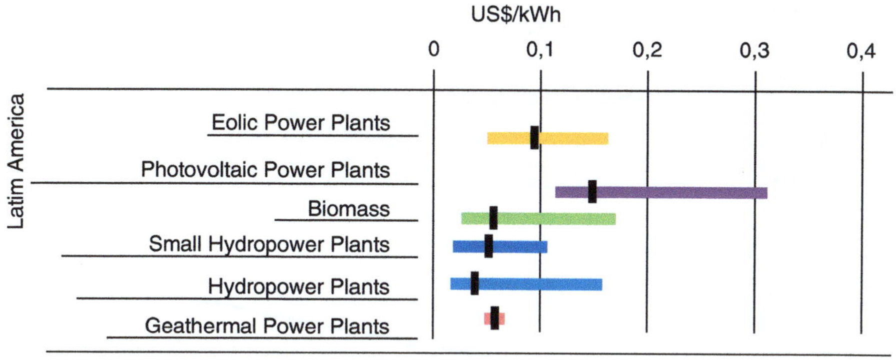

Fig. 4.7 Cost of electricity generation through renewable sources in Latin America in 2012 (Adapted from IRENA 2013)

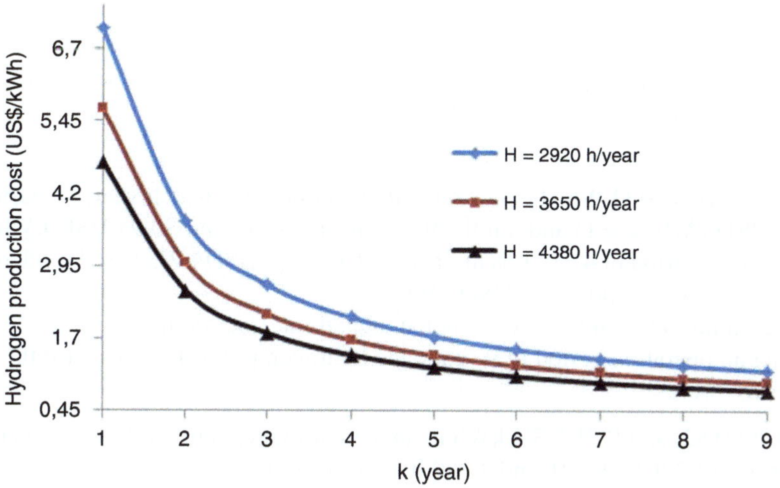

Fig. 4.8 Hydrogen production cost through electrolysis using electricity from wind power plants as a function of *payback* ($r = 8\ \%$)

According to Fig. 4.7, the price of electricity generation from wind power plants is between $0.05 and 0.17 USD/kWh, $0.11 and 0.31 USD/kWh from photovoltaic power plants, and between $0.02 and 0.105 USD/kWh from SHPPs (IRENA 2013).

Given the aforementioned, the following results were obtained:

The charts containing the production costs of 1 Nm^3/h of hydrogen through electrolysis using electricity generated from wind power plants are shown in Figs. 4.8 and 4.9.

According to Fig. 4.8, by setting the annual interest rate at 8 %, it is observed that, for this type of process, the hydrogen production costs for an operation period of 2.920 h/year started at $7.05 USD/kWh ($k = 1$) and, on the 9th year, it was

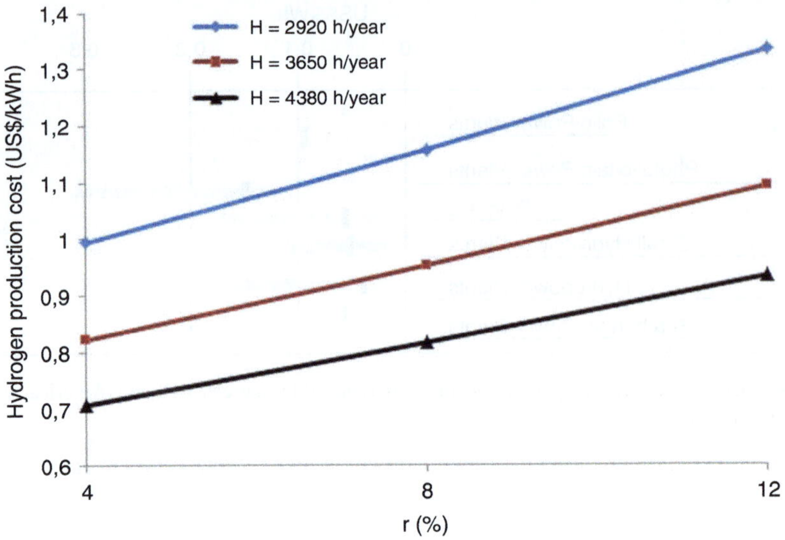

Fig. 4.9 Hydrogen production cost through electrolysis using electricity from wind power plants as a function of annual interest rate ($k = 9$ years)

around $1.16 USD/kWh. For an operation period of 3.650 h/year, it started at $5.67 USD/kWh ($k = 1$) and, on the 9th year, it was around $0.95 USD/kWh. For an operation period of 4.380 h/year, it started at $4.74 USD/kWh ($k = 1$) and, on the 9th year, it was around $0.82 USD/kWh.

According to Fig. 4.9, by varying the annual interest rate in 4, 8, and 12 % and keeping an operation period of 9 years, the hydrogen production costs for this type of process were:

- $0.99; 0.82; and 0.71 USD/kWh at an annual interest rate of 4 % and operation periods of 2.920, 3.650, and 4.380 h/year, respectively
- $1.16; 0.95; and 0.82 USD/kWh at an annual interest rate of 8 % and operation periods of 2.920, 3.650, and 4.380 h/year, respectively
- $1.13; 1.09; and 0.93 USD/kWh at an annual interest rate of 12 % and operation periods of 2.920, 3.650, and 4.380 h/year, respectively

The charts containing the production costs of 1 Nm^3/h of hydrogen through electrolysis using the electricity generated from photovoltaic power plants are shown in Figs. 4.10 and 4.11.

According to Fig. 4.10, by setting the annual interest rate at 8 %, it is observed that, for this type of process, the hydrogen production costs for an operation period of 2.920 h/year started at $7.17 USD/kWh ($k = 1$) and, on the 9th year, it was around $1.78 USD/kWh. For an operation period of 3.650 h/year, it started at $5.79 USD/kWh ($k = 1$) and, on the 9th year, it was around $1.07 USD/kWh. For

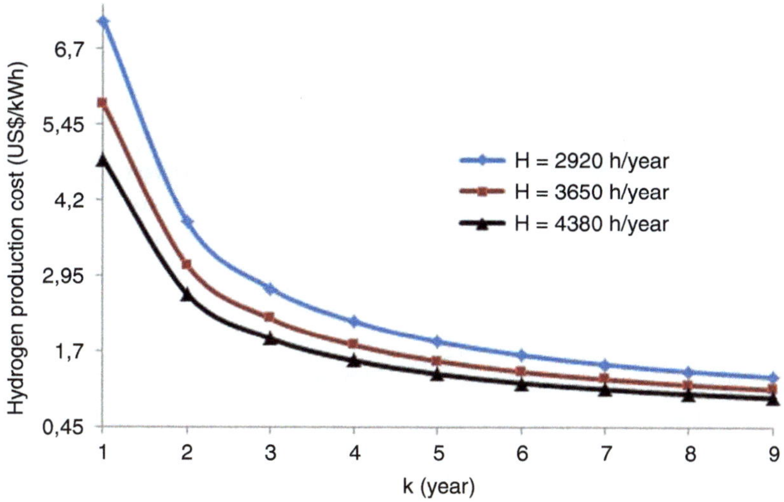

Fig. 4.10 Hydrogen production cost using electricity from photovoltaic power plants as a function of *payback* ($r = 8\,\%$)

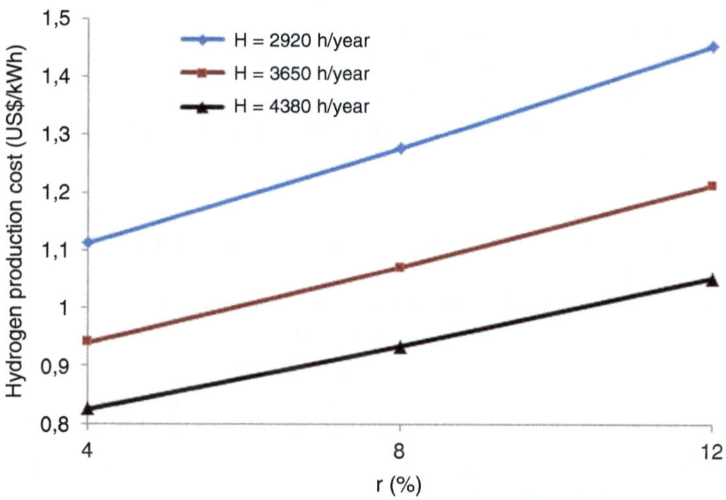

Fig. 4.11 Hydrogen production cost through electrolysis using electricity from photovoltaic power plants as a function of annual interest rate ($k = 9$ years)

an operation period of 4.380 h/year, it started at $4.86 USD/kWh ($k = 1$) and, on the 9th year, it was around $0.93 USD/kWh.

According to Fig. 4.11, by varying the annual interest rate in 4, 8, and 12 % and keeping an operation period of 9 years, the hydrogen production costs for this type of process were:

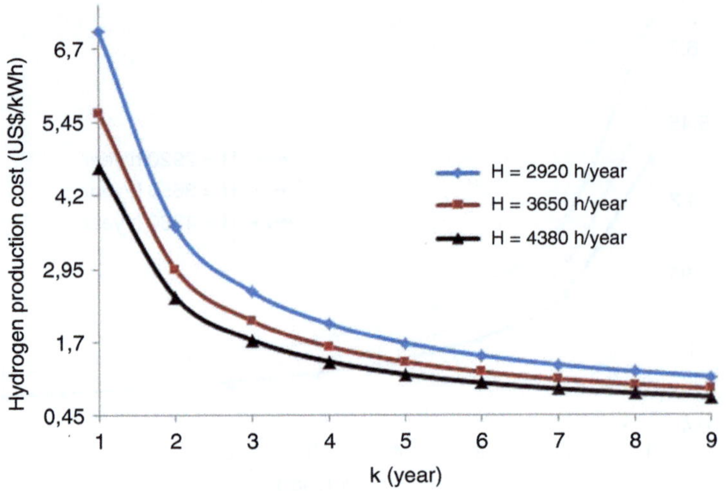

Fig. 4.12 Hydrogen production cost using electricity from hydroelectric power plants as a function of *payback* ($r = 8\ \%$)

- \$1.11; 0.94; and 0.83 USD/kWh at an annual interest rate of 4 % and operation periods of 2.920, 3.650, and 4.380 h/year, respectively
- \$1.28; 1.07; and 0.93 USD/kWh at an annual interest rate of 8 % and operation periods of 2.920, 3.650, and 4.380 h/year, respectively
- \$1.45; 1.21; and 1.05 USD/kWh at an annual interest rate of 12 % and operation periods of 2.920, 3.650, and 4.380 h/year, respectively

The charts containing the production costs of 1 Nm^3/h of hydrogen through electrolysis using the electricity generated from hydroelectric power plants are shown in Figs. 4.12 and 4.13.

According to Fig. 4.12, by setting the annual interest rate at 8 %, it is observed that, for this type of process, the hydrogen production costs for an operation period of 2.920 h/year started at \$0.699 USD/kWh ($k = 1$) and, on the 9th year, it was around \$1.10 USD/kWh. For an operation period of 3.650 h/year, it started at \$5.61 USD/kWh ($k = 1$) and, on the 9th year, it was around \$0.90 USD/kWh. For an operation period of 4.380 h/year, it started at \$4.69 USD/kWh ($k = 1$) and, on the 9th year, it was around \$0.76 USD/kWh.

According to Fig. 4.13, by varying the annual interest rate in 4, 8, and 12 % and keeping an operation period of 9 years, the hydrogen production costs for this type of process were:

- \$0.94; 0.77; and 0.66 USD/kWh at an annual interest rate of 4 % and operation periods of 2.920, 3.650, and 4.380 h/year, respectively
- \$1.10; 0.90; and 0.76 USD/kWh at an annual interest rate of 8 % and operation periods of 2.920, 3.650, and 4.380 h/year, respectively
- \$1.28; 1.04; and 1.88 USD/kWh at an annual interest rate of 12 % and operation periods of 2.920, 3.650, and 4.380 h/year, respectively

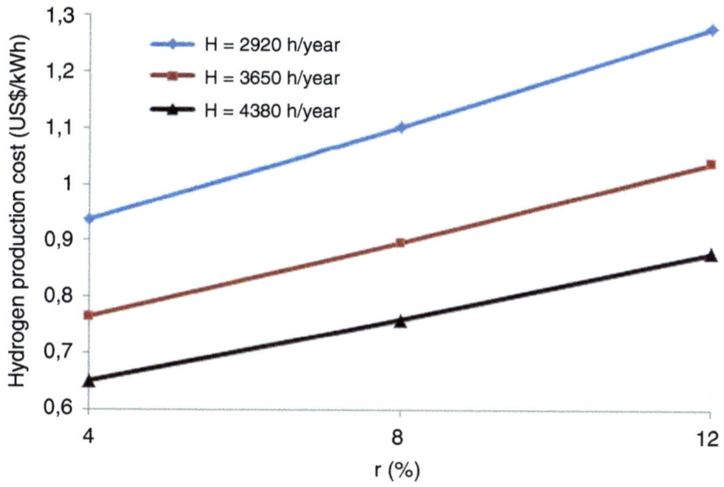

Fig. 4.13 Hydrogen production cost through electrolysis using electricity from hydroelectric power plants as a function of annual interest rate ($k = 9$ years)

4.4 Economic Analysis of Hydrogen Production Process from Algae

For this economic analysis, the following data were considered:

- Every medium contained in reactors 1–5 is aimed at industries that utilize algae (pharmaceutical or biodiesel industries) after the period of hydrogen production
- Area of the reactors for adapting and producing hydrogen: 107.88 m^2 (five reactors being 4.31 m wide and 5 m long)
- Area of the reactors for growing algae: 21.58 m^2 (five reactors being 4.31 m wide and 5 m long). It is considered that the volume contained in 10.79 m^2 of this reactor is transferred to the hydrogen production reactors at every 164 h (adaptation + hydrogen production), and the remaining volume is left in the reactor for algae growth, thus only requiring the medium replenishment (TAP).
- Investment of \$100 USD/m^2 in each closed reactor (Esper et al. 2006; Kunjapur et al. 2010)
- Algae cost: \$10.00 USD (initial cost) (Chlamydomonas Resource Center University of Minnesota 2013)
- Media cost: it was considered the same price for both media, since there is no differential cost when replacing sulfur salts in the first medium with hydrochloric salts in the second medium (Chlamydomonas Resource Center University of Minnesota 2013): \$0.0526 USD/L (James et al. 2009)
- Annual interest rate: 4, 8, and 12 % (BCB 2013)
- Operation period: 2.920, 3.650, and 4.380 h/year of operation (being 8, 10, and 12 h a day for 365 days)

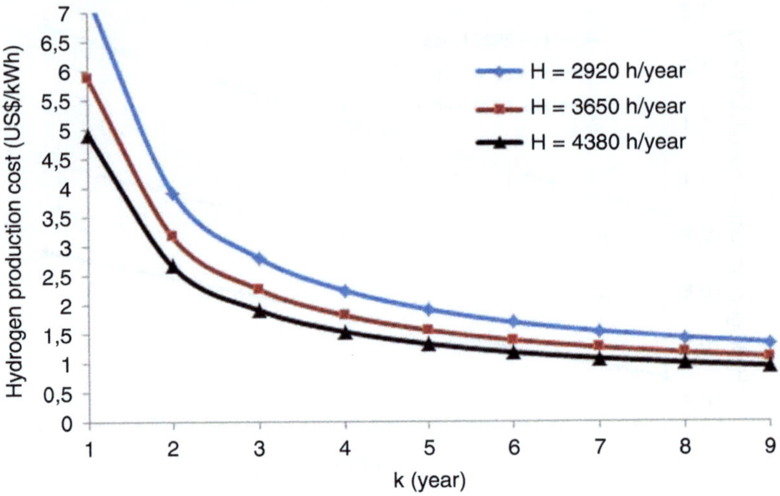

Fig. 4.14 Cost of hydrogen production from algae as a function of *payback* ($r = 8\,\%$)

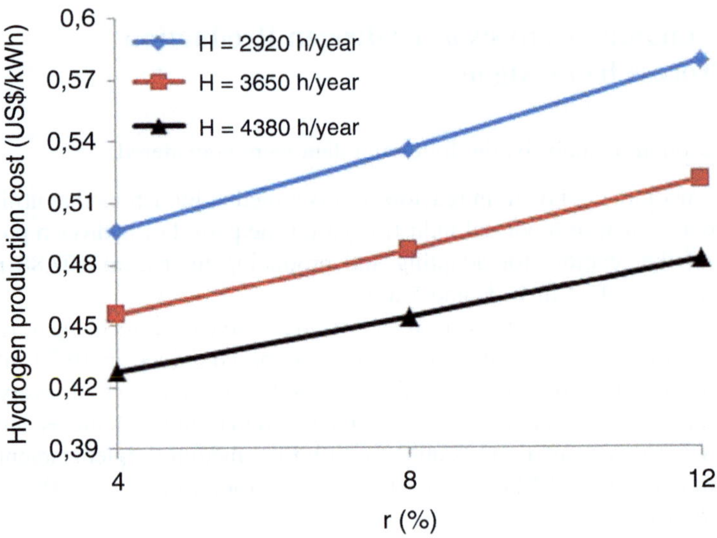

Fig. 4.15 Cost of hydrogen production from algae as a function of annual interest rate ($k = 9$ years)

Given the aforementioned, the following results were obtained:

The charts containing the production costs of 1 Nm^3/h of hydrogen production through the microalga *C. reinhardtii* are shown in Figs. 4.14 and 4.15.

According to Fig. 4.14, by setting the annual interest rate at 8 %, it is observed that, for this type of process, the hydrogen production costs for an operation period of 2.920 h/year started at \$1.96 USD/kWh ($k = 1$) and, on the 9th year, it was around \$0.54 USD/kWh. For an operation period of 3.650 h/year, it started at \$1.62 USD/kWh ($k = 1$) and, on the 9th year, it was around \$0.49 USD/kWh. For an operation period of 4.380 h/year, it started at \$1.40 USD/kWh ($k = 1$) and, on the 9th year, it was around \$0.45 USD/kWh.

According to Fig. 4.15, by varying the annual interest rate in 4, 8, and 12 % and keeping an operation period of 9 years, the hydrogen production costs for this type of process were:

- \$0.50; 0.46; and 0.43 USD/kWh at an annual interest rate of 4 % and operation periods of 2.920, 3.650, and 4.380 h/year, respectively
- \$0.54; 0.49; and 0.45 USD/kWh at an annual interest rate of 8 % and operation periods of 2.920, 3.650, and 4.380 h/year, respectively
- \$0.58; 0.52; and 0.48 USD/kWh at an annual interest rate of 12 % and operation periods of 2.920, 3.650, and 4.380 h/year, respectively

4.5 Economical Comparison Between Hydrogen Production Processes

It was drawn a comparison among the costs of hydrogen production of all studied processes (Fig. 4.16). It was considered an operation period of 3650 h/year, an annual interest rate of 8 % and $k = 9$ years.

According to Fig. 4.16, you can observe that the cost of hydrogen production from electrolysis with photovoltaic energy is the highest and that of the steam reforming of ethanol is the lowest.

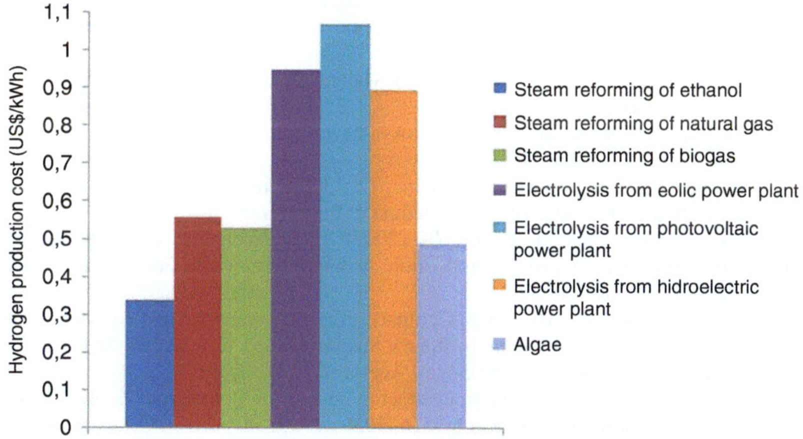

Fig. 4.16 Comparison of the hydrogen production costs of different studied processes

According to the figures presented in the economic analysis, the costs decrease as the capital amortization period and equivalent period of use increase. Investment costs significantly influence the initial and final cost of hydrogen production. Thus, as in the electrolysis processes, the investment in electrolyzers is greater, as the cost of hydrogen produced by this type of process is greater than those using reforming systems, mainly during the first few years of operation.

The influence of operation costs is due to fuel consumption. In the case of the reforming system, the operation cost refers to the vaporization of water in the steam generator and fuel consumption in the reforming reaction. As for the electrolysis, the operation cost refers to the purchase of electricity from power plants to power the electrolyzer and also to the water used in the process. As regards hydrogen generation from algae, the maintenance cost refers to the expenditure on the medium that needs to be replenished, both for algae growth and hydrogen production. It is noted that the initial cost of hydrogen production from algae is lower, but the decrease in cost is less pronounced over time than with the other processes, which is due to the need for replenishing the TAP and TAP-S media constantly. Thus, it is the operation cost of this process which influences the final cost of hydrogen production.

As the investment costs of electrolytic processes are the same, the price difference is quite negligible, which is only due to operation costs. In the case of the reforming processes, the investment in the ethanol reformer is lower than in the biogas and natural gas reformers on account of the price difference to keep the operating temperature difference.

References

BCB-Banco Central do Brasil. http://www.bcb.gov.br/?COPOMJUROS. Accessed 9 Sept 2013

Braga LB, Silveira JL, Silva ME, Tuna CE, Machin EB, Pedroso DT (2013) Hydrogen production by biogas steam reforming: a technical, economic and ecological analysis. Renew Sustain Energy Rev 28:166–173

Chlamydomonas Resource Center. University of Minnesota. http://chlamycollection.org/media/. Accessed 16 Apr 2013

Esper B, Badura A, Rogner M (2006) Photosynthesis as a power supply for bio-hydrogen production. Trends Plant Sci 11:543–549

IRENA—International Renewable Energy Agency (2013) Summary for policy makers: renewable power generation costs—Abu Dhabi, United Arab Emirates, 12 p

James BD, Baum GM, Perez J, Baum NK (2009) Technoeconomic Boundary Analysis of Biological Pathways to Hydrogen Production. National Renewable Energy Laboratory, Virginia, p 207, NREL/SR-560-46674

Kothari R, Buddhi D, Sawhney RL (2008) Comparison of environmental and economic aspects of various hydrogen production methods. Renew Sustain Energy Rev 12:553–563

Krona. http://www.krona.srv.br/display05.htm. Accessed 10 Oct 2012

Kunjapur AM, Eldridge RB (2010) Photobioreactor Design for Commercial Biofuel Production from Microalgae. Ind Eng Chem Res 8:3516–3626

Lora EES, Nascimento MAR (2004) Geração Termelétrica: Planejamento, Projeto e Operação, vol 2, Rio de Janeiro. In: Interciência

Nitidor. http://www.h2nitidor.it/h2n/. Accessed 27 Apr 2013

Saur G (2008) Wind-to-hydrogen project: electrolyzer capital cost study. National Renewable Energy Laboratory, Colorado, 48 p (NREL/TP-550-44103)

Silva ME (2010) Análise experimental da reforma a vapor de etanol: aspectos técnicos, econômicos e ecológicos. Tese (Doutorado)-Curso em Engenharia Mecânica. Faculdade de Engenharia de Guaratinguetá, Universidade Estadual Paulista, Guaratinguetá

Silveira JL, Gomes LA (1999) Fuel cell cogeneration system: a case of techno economic analysis. Renew Sustain Energy Rev 3:233–242

Webcalendar (2014) http://www.web-calendar.org/pt/world/south-america/brazil/brasilia--07?menu=sun. Accessed 28 May 2014

Chapter 5
Ecological Efficiency of Some Hydrogen Production Processes

Lúcia Bollini Braga, Celso Eduardo Tuna, Fernando Henrique Mayworm de Araujo, Lucas Fachini Vane, Daniel Travieso Pedroso, and José Luz Silveira

5.1 Introduction

Pollutant gases emission into the atmosphere directly harms human, animal and vegetable health. Among these, the most harmful gases are carbon monoxide (CO), nitrous oxides (NOx), sulphur dioxide (SO_2) and particulate matter (MP). Carbon dioxide (CO_2), although not being harmful to human health, is one of the most significant contributors to the greenhouse effect increase, due to being able to retain infrared radiation; SO_2 stands out as one of the main causes of acid rain, and particulates are the components that lead to a decrease in the amount of radiation that reaches the Earth's surface (Costa 2011).

Many definitions of equivalent carbon dioxide can be found in literature. One of them was defined by Cardu and Baica (1999) who introduced this concept to calculate ecological efficiency.

Ecological efficiency considers the emissions from fuel and electricity used in a given system, allowing the quantification of their pollution level. It ranges from 0 to 1: a value of 0 means 100 % environmental impact, thus indicating a very polluting system, and 1 means 0 % environmental impact, i.e. a non-polluting system. The concept of equivalent carbon dioxide [$(CO_2)e$] was based on the maximum allowable concentration of CO_2 in the atmosphere (Júnior 2004). This coefficient, whose

L.B. Braga • C.E. Tuna • F.H.M. de Araujo • L.F. Vane • D.T. Pedroso
Group of Optimization of Energy Systems—GOSE, College of Engineering of Guaratinguetá, Institute of Bioenergy Research—IPBEN, São Paulo State University—UNESP, Dr. Ariberto Pereira da Cunha Ave, 333, Guaratinguetá 12516-410, SP, Brazil

J.L. Silveira (✉)
Group of Optimization of Energy Systems—GOSE, College of Engineering of Guaratinguetá, Institute of Bioenergy Research—IPBEN, São Paulo State University—UNESP, Dr. Ariberto Pereira da Cunha Ave, 333. Pedregulho, Guaratinguetá 12516-410, São Paulo, Brazil
e-mail: joseluz@feg.unesp.br

J.L. Silveira (ed.), *Sustainable Hydrogen Production Processes*, Green Energy and Technology, DOI 10.1007/978-3-319-41616-8_5

unit is in kg of emissions per kg of fuel ($kg(CO_2)_e/kg_{fuel}$), is calculated according to Eq. (5.1).

$$(CO_2)e = CO_2 + 80(SO_2) + 50(NO_x) + 67(MP) \tag{5.1}$$

In order to quantify the environmental impact, Cardu and Baica (1999) proposed an indicator, defined as the ratio between the fuel's equivalent carbon dioxide and the energy contained in the fuel being used (Qi). This indicator was named Pollutant Indicator, represented by Π_g, as Eq. (5.2).

$$\Pi_g = \frac{(CO_2)e}{Qi} \tag{5.2}$$

where:

$(CO_2)e$—Equivalent carbon dioxide (kg/kg fuel)
Qi—Calorific power of the fuel (LHV) (MJ/kg fuel)
Π_g—Pollutant Indicator (kg/MJ)

By considering the hypothesis that the best fuel is the one which has the lowest Pollutant Indicator value, it was proposed a more complex and dimensionless indicator expressing the ecological impact of harmful gases that are released into the atmosphere and produced by a power plant. The proposed indicator is called as ecological efficiency (ε), which is given by Eq. (5.3).

$$\varepsilon = \left[\frac{0.204 \cdot \eta_{system} \cdot \ln(135 - \Pi_g)}{\eta_{system} + \Pi_g}\right]^{0.5} \tag{5.3}$$

In the case of hydrogen production processes, the same equation cannot be used (5.3), therefore the ecological efficiency equation defined by Cardu and Baica (1999) had to be adapted.

For electrolysis processes that do not utilize energy from fuels, but the electricity from wind, photovoltaic and hydroelectric power plants, it was considered the GWP (Global Warming Potential) Pollutant Indicator. It is used in the LCA (Life Cycle Assessment) which quantifies the environmental impact of the development and implementation of a technology, and such indicator's unit is $kg(CO_{2(eq)})/kWh$. This indicator was adapted for hydrogen production and its unit started being referred as kg of emission per kg of hydrogen production (kg $(CO_2)e/kgH_2$).

In all processes, the emissions under consideration are those concerning the form of obtaining the fuels to be used in the hydrogen production equipment and the emissions during the process itself. Those generated from the development of devices were not considered, i.e. the reformers, electrolyzers and photobioreactors.

The ecological efficiency equation to be used for hydrogen production processes is presented in Eq. (5.4). Appendix B shows the development and limit conditions adopted for obtaining this equation.

$$\varepsilon = \left[\frac{0.25 \cdot \eta_{\text{system}} \cdot \ln(51 - indicator)}{\eta_{\text{system}} + indicator}\right]^{0.023} \tag{5.4}$$

where:

η_{system}—thermodynamic efficiency of the process [%]
Indicator—system's Pollutant Indicator [kg$(CO_2)_e$/kWh]

5.2 Ecological Efficiency of Steam Reforming Processes

The calculation of ecological efficiencies of steam reform processes was based on Fig. 5.1.

The reforming process consists of a steam generator that produces steam for the process and a system composed of a reformer and a water-gas *shift* reactor. Hydrogen is considered as the only energy product of the overall process, and the fuel used in the steam generator and the reforming process is regarded as reactant.

5.2.1 Steam Reforming Process of Ethanol

In the steam reforming process of ethanol, bagasse was chosen as fuel to be burned in the steam generator, and ethanol was selected to be consumed in the steam reforming reaction. The stoichiometric equation of sugar cane bagasse combustion and the overall steam reforming equation were used in the calculations. It was considered 85 % efficiency of combustion in the steam generator (Lora and Nascimento 2004), 90.78 % efficiency of the reforming reaction and 73 % efficiency of the overall process. The bagasse composition on a dried basis is presented in Table 5.1.

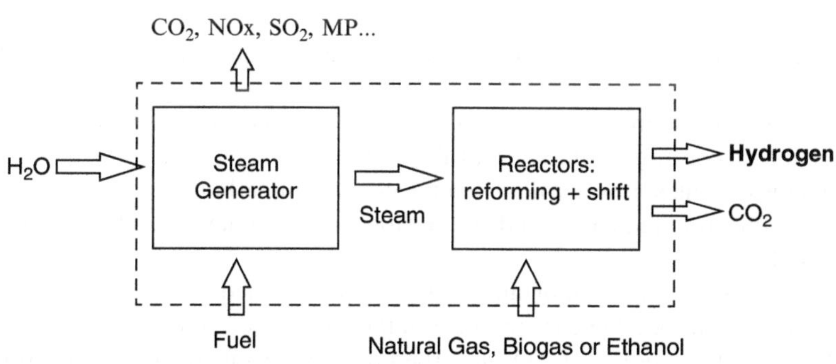

Fig. 5.1 Steam reforming processes flow chart

Table 5.1 Elemental composition of bagasse on a dried basis (Adapted from Jenkins et al. 1998)

Component	C	H	O	N	S	Cl	Ash	Total
% in mass (dry basis)	48.64	5.87	42.82	0.16	0.04	0.03	2.44	100

Table 5.2 Composition estimated for atomic weight and number of mol of each of the components of bagasse

Component	C	H	O	N	S	Cl	Ashes
Mass (kg) (50 % humidity)	24.32	2.94	21.41	0.08	0.02	0.02	1.22
Atomic weight (g)	12	1	16	14	32	35.5	–
Number of moles	2.03	8.49	4.12	0.006	0.0006	0.0004	–

Table 5.3 Process emissions from hydrogen production through the steam reforming of ethanol

Components	Sugarcane bagasse combustion (kg CO_2/kg BC)	Steam reforming of ethanol (kg CO_2/kg Et.)
CO_2	0.76	1.73

As in the burning process of bagasse in the steam generator, it was not considered any previous drying procedure, but that the bagasse enters the steam generator with about 50 % humidity. Thus, the chemical composition, atomic weight and number of moles of each of the components of bagasse are shown in Table 5.2 by considering 100 kg of it.

As Table 5.2, the stoichiometric equation of bagasse is presented in Eq. (5.5) and the reforming reaction in Eq. (5.6).

$$\begin{aligned} &2.03C + 8.49H + 4.12O + x(\alpha O_2 + 3.7\alpha N_2) \\ &\rightarrow 2.03CO_2 + 4.25H_2O + 3.7x\alpha N_2 + x(\alpha - 1)O_2 \end{aligned} \tag{5.5}$$

where:

$\alpha = 1.20$ (considering 20 % air excess)

$$C_2H_6O + 3H_2O \rightarrow 2CO_2 + 6H_2 \tag{5.6}$$

The emissions, according to the stoichiometric equations, are indicated in Table 5.3.

It was also considered the emissions of NOx, SO_2 and MP of sugarcane bagasse combustion, according to literature (Table 5.4).

From these values, it was calculated the equivalent carbon dioxide [$(CO_2)_e$], the pollutant indicator (indicator) and the ecological efficiency (ε). These values are shown in Table 5.5.

Given the fact that ethanol and bagasse are obtained from sugar cane, it is possible to incorporate the carbon cycle in the calculation of ecological efficiency by subtracting the amount of CO_2 absorbed during the sugarcane growth from the equivalent carbon dioxide, as shown in Fig. 5.2.

Table 5.4 Steam generator emissions according to literature data (Reproduced from Lora and Nascimento 2004)

Components	Sugarcane bagasse combustion (kg/kg SB)
SO_2	–
NOx	$1.2 \cdot 10^{-3}$
MP	$7.1 \cdot 10^{-3}$

Table 5.5 Ecological analysis results for the steam reforming of ethanol

Indicators	Values
$(CO_2)_e$	3.56 kg $(CO_2)_e$/kg fuel
Indicator	39 kg $(CO_2)_e$/kgH_2
ε	90 %

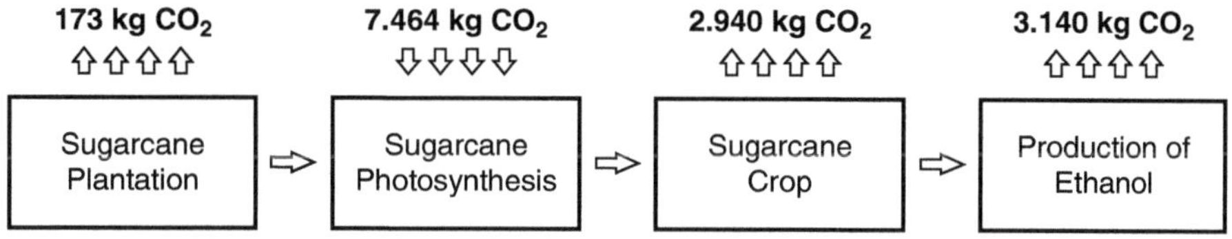

Fig. 5.2 Carbon dioxide cycle for 1000 L of produced ethanol (Adapted from Silveira et al. 2009)

Table 5.6 Ecological analysis results for the steam reforming of ethanol considering the carbon cycle

Indicators	Values
$(CO_2)_e$	2.03 kg $(CO_2)_e$/kg fuel
Π_g	21 kg (CO_2)e/kgH_2
ε	92.13 %

For every 1000 L of ethanol being produced, 7464 kg of carbon dioxide (*photosynthesis*) are consumed and 6253 kg are released, showing a positive balance of 1211 kg of carbon dioxide absorption. Thus, once considering this parameter and knowing that 1 ton of sugarcane generates 83.33 L of ethanol and 250 kg of bagasse, it is obtained new equivalent carbon dioxide [$(CO_2)_e$], pollutant indicator and ecological efficiency (ε) values, all introduced in Table 5.6.

5.2.2 *Steam Reforming Process of Natural Gas*

Natural gas was chosen as fuel to be burned in the steam generator and consumed in the reaction for the steam reforming process of natural gas. The stoichiometric equation of the steam generator for this type of fuel is shown in Eq. (5.7) and the overall steam reforming reaction Eq. (5.8) was used for such calculations.

Table 5.7 Process emissions from hydrogen production through the steam reforming of natural gas

Components	Natural gas combustion (kg CO_2/kg NG)	Steam reforming of natural gas (kg CO_2/kg NG)
CO_2	2.50	2.66

Table 5.8 Steam generator emissions according to literature data (Reproduced from Lora and Nascimento 2004)

Components	Sugarcane bagasse combustion (kg/kg)
SO_2	–
NOx	$3.47 \cdot 10^{-3}$
MP (average)	$6.13 \cdot 10^{-5}$

Table 5.9 Ecological analysis results for the steam reforming of natural gas

Indicators	Values
$(CO_2)_e$	12.84 kg (CO_2)e/kg fuel
Indicator	18.37 kg(CO_2)e/kgH_2
ε	92.41 %

Table 5.10 Ecological analysis results considering the emissions from the natural gas production process

Indicators	Values
$(CO_2)_e$	7.52 kg/kg fuel
Indicator	25.05 kg(CO_2)e/kgH_2
ε	91.63 %

$$0.9CH_4 + 0.1C_2H_6 + x(\alpha O_2 + 3.7\alpha N_2) \rightarrow 1.1CO_2 + 2.1H_2O + 3.7x\alpha N_2 + x(\alpha - 1)O_2 \quad (5.7)$$

$$0.9CH_4 + 0.1C_2H_6 + 2.2H_2O \rightarrow 1.1CO_2 + 4.3H_2 \quad (5.8)$$

For the steam generator combustion, it was considered 20 % air excess ($\alpha = 1.20$) and 90 % efficiency (Lora and Nascimento 2004); it was also considered 79.05 % efficiency for the reforming and 71.78 % efficiency for the overall process.

The emissions, according to the stoichiometric equations, are indicated in Table 5.7.

It was also considered, based on literature, the emissions of NOx, SO_2 and MP on the combustion of bagasse, as shown in Table 5.8.

From these values, it was calculated the equivalent carbon dioxide [$(CO_2)_e$], the pollutant indicator (indicator) and the ecological efficiency (ε). These values are shown in Table 5.9.

As in the case of ethanol, it was considered its obtainment through sugarcane; as for the natural gas case, it was taken into account their emission in the production process. According to Spath and Mann (2000), the $(CO_2)_e$ of such a process is 1.781 kg CO_2/kg NG, and thus new equivalent carbon dioxide [$(CO_2)_e$], pollutant indicator (indicator) and eco-efficiency (ε) values were obtained, as shown in Table 5.10.

5.2.3 Steam Reforming Process of Biogas

Biogas was chosen as fuel to be burned in the steam generator and consumed in the reforming reaction for the steam reforming process of biogas. The stoichiometric equation of the steam generator for this type of fuel is shown in Eq. (5.9), and the overall steam reforming reaction Eq. (5.10) was used for such calculations.

For the steam generator combustion, it was considered 20 % air excess ($\alpha = 1.20$) and 90 % efficiency (Lora and Nascimento 2004); it was considered 74.4 % efficiency of the steam reforming and 68.4 % efficiency for the overall process.

$$0.757CH_4 + 0.243CO_2 + x(\alpha O_2 + 3.7\alpha N_2) \rightarrow 1CO_2 + 1.514H_2O + 3.7x\alpha N_2 + x(\alpha - 1)O_2 \tag{5.9}$$

$$CH_4 + 2H_2O \rightarrow 4H_2 + CO_2 \tag{5.10}$$

The emissions, according to the stoichiometric equations, are indicated in Table 5.11.

It was also considered, based on literature, the emissions of NOx, and MP on the combustion of biogas, as shown in Table 5.12.

From these values, it was calculated the equivalent carbon dioxide [$(CO_2)_e$], the pollutant indicator (indicator) and the ecological efficiency (ε), as shown in Table 5.13.

In the case of the ecological efficiency of this type of production process, the value of equivalent CO_2 emissions that occur during the biogas production from

Table 5.11 Process emissions from hydrogen production through the steam reforming of biogas

Components	Biogas combustion (kg CO_2/kg BG)	Steam reforming of biogas (kg CO_2/ kg BG)
CO_2	1.74	1.07

Table 5.12 Steam generator emissions while burning biogas according to literature data (Adapted from the U.S. Environmental Protection Agency EUA 1995; Beylot et al. 2012)

Components	Biogas combustion (kg/kg BG)
SO_2	–
NOx	$7.36 \cdot 10^{-4}$
MP	$1.80 \cdot 10^{-4}$

Table 5.13 Ecological analysis results of the steam reforming of biogas

Indicators	Values
$(CO_2)_e$	2.87 kg (CO_2)e/kg fuel
Indicator	15.43 kg(CO_2)e/kgH_2
ε	92.74 %

Table 5.14 Emissions of the biogas production process from manure in [kg(CO_2)/kg manure] (Adapted from Poeschl et al. 2012)

CH_4	SO_2	NOx	MP
$4 \cdot 10^{-4}$	$6.6 \cdot 10^{-6}$	$9.7 \cdot 10^{-6}$	$4.5 \cdot 10^{-6}$

Table 5.15 Ecological analysis results of the steam reforming of biogas with emissions from its production process

Indicator	Values
$(CO_2)_e$	3.14 kg(CO_2)e/kg biogas
Π_g	16.87 kg(CO_2)e/kgH_2
ε	92.54 %

manure must be added. These emissions are presented in Table 5.14 according to literature data.

By considering the ratio of 0.037 kg/biogas to kg/manure (Coelho et al. 2008), new (CO_2)e, pollutant indicator and ecological efficiency values were obtained, as shown in Table 5.15.

In the case of biogas, it is also relevant to include the amount of emissions that would not be released into the environment in the ecological calculation, as regards the use of manure. The greatest problem of manure being exposed to the environment is that the emission of CH_4, as aforementioned, is 21 times more harmful to the environment than that of CO_2.

According to Govoni (2011), there are techniques to calculate the amount of methane emitted by each type of animal waste. In the case of manure, 0.13 m^3 of methane is emitted per every kg of manure. Thus, by taking into account the 0.037 kg/biogas:kg/manure ratio (Coelho et al. 2008) and multiplying this value by 21, it results in the amount of 48.51 kg(CO_2)e that would be released into the atmosphere. Based on this parameter, the (CO_2)e value would be negative, therefore indicating a carbon credit, thus the ecological efficiency of the process as a whole would be 100 %.

5.3 Ecological Efficiency of Electrolysis

In the case of the calculation of ecological parameters of hydrogen production processes from electrolytic processes using electricity from renewable sources, it was considered the emissions from manufacturing the devices used in wind and photovoltaic power plants and, in the case of hydroelectric power plants, the emissions from its development and flooded area. As previously described in item 5.1, in the case of calculating the ecological efficiency of these processes, the formula from Cardu and Baica (1999) was adapted and used as a function of the pollutant indicator, which is shown in Appendix B.

5.3.1 Eolic Energy

According to literature, a wind turbine's GWP is 15.3 $kgCO_{2(e)}/MWh$ (Turconi et al. 2013) and, by converting this value, it is obtained a new pollutant indicator of 0.60 $kg\,CO_{2(e)}/kg\,H_2$. The thermodynamic efficiency of this type of hydrogen production process is 29.5 % and an ecological efficiency of 94.9 % is achieved.

5.3.2 Photovoltaic Energy

According to literature, a PV panel's GWP is 52 $kg\,CO_{2(e)}/MWh$ (Turconi et al. 2013) and, by converting this value, it is obtained a new Pollutant Indicator of 2.05 $kg\,CO_{2(e)}/kg\,H_2$. The thermodynamic efficiency of this type of hydrogen production process is 11.8 % and an ecological efficiency of 93.46 % is achieved.

5.3.3 Hydroelectric Generation from Water Spillage

In the study of the ecological efficiency of electrolysis in hydroelectric power plants, it was considered the emissions from plants with a reservoir (emission of CH_4 and CO_2 as a function of km_2 flooded and depth). According to literature, this system's GWP is 8.8 $kg\,CO_{2(e)}/MWh$ (Turconi et al. 2013) and, by converting this value, it is obtained a new pollutant indicator of 0.34 $kg\,CO_{2(e)}/kg\,H_2$. The thermodynamic efficiency of this type of hydrogen production process is 78 % and an ecological efficiency of 99 % is achieved.

5.4 Ecological Efficiency of Hydrogen Production from Microalgae

For the calculation of (CO_2)e as regards the alga *C. reinhardtii*, it was considered its capacity to absorb carbon dioxide during its growth period. For such a purpose, a compressor for reducing the volume of CO_2 in the environment was attached to the growth tank. Equation (5.11) shows the mechanism of CO_2 fixation through the period of algae photosynthesis (Kothari et al. 2008):

$$CO_2 + H_2O + \text{solar energy} \rightarrow \frac{1}{6} C_6H_{12}O_6 + O_2 \tag{5.11}$$

During this phase, the maximum amount of CO_2 fixed by this type of alga is 1.01 kg/m_3 day (Zeng et al. 2011; Skjanes et al. 2007). By considering the volume which had already been calculated previously for the growth tank, it was found operation periods of 2.920, 3.650 and 4.380 h/year, where 72, 90 and 108 kg CO_2/year were fixed, respectively.

As the emissions during this process are minimal (only from the energy consumption of the system's two pumps and compressor) and that there is high CO_2 consumption by algae, the process's ecological efficiency can be considered as being approximately 100 %.

5.5 Ecologic Comparison Between the Studied Hydrogen Production Processes

The values of ecological efficiency of hydrogen production processes can be observed in Table 5.16. Only two hydrogen production processes obtained 100 % efficiency, which are: the photobiological hydrogen production from algae and the steam reforming of biogas, taking into account the emissions that would no longer be released into the environment. Subsequently, the best ecological efficiencies are obtained from the electrolysis using electricity from hydroelectric power plants with 99 %, electrolysis using electricity from wind farms with 94.9 %, steam reforming of ethanol considering the sugarcane cycle, and the steam reforming of natural gas given the emissions in its obtainment.

Hydrogen production processes through electrolysis and algae growth do not present an equivalent carbon dioxide value because these do not use polluting fuels, but only green energy-generating sources (wind, sunlight and water). Among these processes, as for electrolytic processes, emissions from the development of power-generating devices were considered (wind turbines, photovoltaic panels and hydro-electric power plants), therefore presenting Pollutant Indicator values, however, in

Table 5.16 Ecological efficiencies of hydrogen production processes

Processes	$(CO_2)e$ [kg $CO_{2(e)}$/kg fuel]	Indicator [kg $CO_{2(e)}$/kg H_2]	ε [%]
Steam reforming of ethanol	3.56	39	90
Steam reforming of ethanol[a]	2.03	21	92.13
Steam reforming of natural gas[b]	7.52	25.05	91.63
Steam reforming of biogas	3.14	16.87	92.54
Steam reforming of biogas[c]	–	–	100
Electrolysis using wind farm electricity	–	0.60	94.9
Electrolysis using PVP plant electricity	–	2.05	93.46
Electrolysis using electricity from HPP with reservoir	–	0.34	99
Algae	–	–	100

[a]Considering the carbon cycle
[b]Considering its obtainment
[c]Considering the amount of methane that is no longer emitted

the case of algae, no emission was considered, but only the amount of CO_2 that this alga absorbs during its photosynthesis process.

Despite the fact that the Pollutant Indicators present significant differences among themselves, the ecological efficiency does not vary in the same proportion because it is influenced by the thermodynamic efficiency of the processes and the LHV of fuels.

References

Beylot A, Villeneuve J, Bellenfant G (2012) Life Cycle Assessment of landfill biogas management: sensitivity to diffuse and combustion air emissions. Waste Manag 33:401–411

Cardu M, Baica M (1999) Regarding a global methodology to estimate the energy-ecologic efficiency of thermopower plants. Energy Convers Manag 40:71–87

Coelho ST, Monteiro MB, Ghilardi A, Karniol MR (2008) Atlas de Bionergia do Brasil. Projeto Fortalecimento Institucional do CENBIO

Costa AAS (2011) Caracterização de Motor Aeronáutico Utilizando Misturas de Gasolina de Aviação e Etanol: Aspectos Fluidodinâmicos, Termodinâmicos e Ecológicos, 179 f. Teses (Doutorado)—Curso de Engenharia Mecânica, Departamento de Energia, Faculdade de Engenharia de Guaratinguetá, Guaratinguetá

Govoni JB (2011) Análise da viabilidade técnica econômica da utilização do biogás proveniente de resíduos de grandes granjas, 89 f. Dissertation. Curso de Engenharia Mecânica, Departamento de Energia, Faculdade de Engenharia de Guaratinguetá, Guaratinguetá

Jenkins BM, Baxter LL, Miles TR Jr, Miles TR (1998) Combustion properties of biomass. Fuel Process Technol 54:17–46

Júnior ACF (2004) Análise da viabilidade econômica da produção de hidrogênio em usinas hidrelétricas: estudo de caso em Itaipu, 143 f. Dissertation, Curso de Engenharia de Produção, Departamento de Produção, Universidade Federal de Santa Catarina, Florianópolis

Kothari R, Buddhi D, Sawhney RL (2008) Comparison of environmental and economic aspects of various hydrogen production methods. Renew Sustain Energy Rev 12:553–563

Lora EES, Nascimento MAR (2004) Geração Termelétrica: Planejamento, Projeto e Operação, vol 2. Interciência, Rio de Janeiro

Poeschl M, Ward S, Owende P (2012) Environmental impacts of biogas deployment e Part I: life cycle inventory for evaluation of production process emissions to air. J Cleaner Prod 24:168–183

Silveira JL, Braga LB, Souza ACC, Antunes JS, Zanzi R (2009) The benefits of ethanol use for hydrogen production in urban transportation. Renew Sustain Energy Rev 13:2525–2534

Skjanes K, Lindblad P, Muller J (2007) Bio-CO_2—a multidisciplinary, biological approach using energy. Biomol Eng 24:405–413

Spath PL, Mann MK (2000) Life cycle assessment of a natural gas combined-cycle power generation system. NREL/TP-570-27715. U.S. Department Of Energy, Colorado

Turconi R, Boldrin A, Astrupe T (2013) Life Cycle Assessment (LCA) of electricity generation technologies: overview, comparability and limitations. Renew Sustain Energy Rev 28:555–565

U.S. Environmental Protection Agency (EUA) (1995) Compilation of air pollutant emission factors, 5th edn. Washington, DC, 26 p

Zeng X, Danquah MK, Chen XD, Lu Y (2011) Microalgae bioengineering: from CO_2 fixation to biofuel production. Renew Sustain Energy Rev 15:3252–3260

Chapter 6
Sustainability Assessment of Hydrogen Production Techniques in Brazil: A Multi-criteria Analysis

Luis Carlos Félix Tapia, Rolando Zanzi Vigouroux, and José Luz Silveira

Throughout this discussion, a sustainability assessment of hydrogen production technologies in Brazil is conducted through the multi-criteria analysis. After defining an initial framework for decision-making, options for hydrogen production were reviewed and selected. These were evaluated and weighted against selected sustainability indicators that fitted the established framework within a weighting matrix. An overall score was obtained after the assessment which ranked hydrogen production techniques based on renewable energy sources, which ended up in first place. The final scoring of options was analyzed, and it was concluded that several approaches could be adopted for interpreting results and their further integration into the policy-making. After concluding that the selection of the right approach depends on the time scale targeted for implementation amongst other multi-disciplinary factors, the use of MCA as an evaluation tool along with overarching sustainability indicators can aid in narrowing uncertainties and providing a clear understanding of the variables surrounding the problem at hand.

L.C.F. Tapia • R. Zanzi Vigouroux (✉)
Department of Industrial Ecology, Royal Institute of Technology (KTH), Teknikringen 34 S, Stockholm 100 44, Sweden
e-mail: rolando@kth.se

J.L. Silveira (✉)
Group of Optimization of Energy Systems—GOSE, College of Engineering of Guaratinguetá, Institute of Bioenergy Research—IPBEN, São Paulo State University—UNESP, Dr. Ariberto Pereira da Cunha Ave, 333. Pedregulho, Guaratinguetá 12516-410, São Paulo, Brazil
e-mail: joseluz@feg.unesp.br

J.L. Silveira (ed.), *Sustainable Hydrogen Production Processes*, Green Energy and Technology, DOI 10.1007/978-3-319-41616-8_6

6.1 Introduction

Repetitive attempts to lobby sustainability and protect natural resources along with the constitutional safeguard of society have made political institutions clash, as interests of separate wings conflict with each other. Political wars loom particularly in countries where inequality is high and the division of social classes remains steeply marked. This fact has sometimes created grudge between social levels that depend on natural resources for subsistence (i.e., indigenous populations) and those attempting to exploit natural resources for generating profit and who usually have access to greater political power.

Previous events that include the political uprise of environmentally or socially concerned individuals have led to the identification of key stakeholders that take part within the sustainability agenda, representing both ends. These stakeholders not only challenge the disproportioned growth of multinational companies or governments, but also create a benchmark on social awareness and a pathway for action (The Guardian 2013a). A prime example is the case of the newly established political party "Sustainability Network" in Brazil represented (led) by Marina Silva, the politician and former Chico Mendez colleague, at the beginning of 2013. Although the newly formed party will likely take social equality and environmental issues as priority within its political agenda, it is important to acknowledge the reasons why other stakeholder groups have supported Mrs. Silva with a way to assemble the party and focus on striving towards sustainability (BBC News 2013).

Despite the fact that division may exist within political wings, decision-making is still required for policy-making, which drives further development of countries and cultures. Particularly in situations where sustainability is the main component of a program or policy, it becomes important that suitable indicators are available for a proper evaluation of projects and matters that may arouse controversy. Providing poor quality indicators to policy-makers can prove challenging to the point of backfire or even social catastrophe.

Such is the case of the Belo Monte hydroelectric power plant project in Brazil, where indigenous populations were severely affected by their displacement due to construction of massive dams and eventual flooding of indigenous settlement areas (The Guardian 2013b). The Belo Monte dam, one of the largest projects in Brazil, has been given green light to proceed with its construction. It was later found that the Environmental Impact Assessment for the project remained incomplete. A supreme court ruled swiftly in issuing a halt for the project, delaying its commencement due to unsuccessful negotiations to relocate 20,000 indigenous individuals.

Although a resolution for the Belo Monte issue still lies in limbo, the paradox of developing important national projects without adequate social and environmental indicators can influence policies that appear to be created for the benefit of all involved levels of society, while, in fact, other sectors of the social strata will become highly impoverished or impacted. In cases like the Belo Monte project, it is critical to account for all involved stakeholders while developing indicators, as they become the main tools to create required legislation for stakeholder protection. Distinguishing the different sustainability dimensions and enabling stakeholders to represent such dimensions as a part or a whole can elucidate the way to create new sustainability indicators or improve the existing ones. By doing so, policy-makers would then make informed-enhanced decisions, translating into actions that would be adjusted more efficiently to the everyday changing aspects of society.

The need for sustainable indicators that are able to portray the current situation of any given system around the globe and accurately predict environmental and economic development or impacts in any time increment in the future can become challenging, or perhaps impossible to be accomplished. Some studies have concluded that "no set of indicators are universally accepted, backed by compelling theory, rigorous data collection and analysis, and influential in policy" (Parris et al. 2003). Based on the aforementioned assumption, what is left then is to modify the existing indicators and adapt them accordingly into a targeted decision-making context.

By molding sustainability indicators into a specific decision context, decision-makers could potentially solve existing social issues that now restrict populations from proper development. Nowadays, one of the most pressing issues that will have a great impact in the future is the increasing demand for energy resources. This issue has placed a heavy burden on governments around the world, particularly on developing countries, putting great strains towards global climate, food security, and social development.

Continuous demand for energy sources at a global scale to satisfy increasing population numbers and mass exodus of people from rural to metropolitan areas has reached alarming rates within the past years, and it is expected to increase even further by the year 2050. While international discussions take place with regard to peak energy resources and upcoming decrease in the production of such, the outlook for alternate energy sources that have a minimal environmental impact and are economically and technically feasible have become the focal point for both developed and developing nations.

Global sustainable development requires a supply of clean and affordable energy sources that avoid or minimize social and environmental impacts. Since all current energy sources may lead to a few environmental impacts, increasing the efficiency of known power generation and transport technologies can alleviate concerns regarding greenhouse gas emissions and their impact on the world's climate (Dincer 2006). Increasing efficiencies, however, will not be enough to entirely reduce the stress caused by environmental damage and social dilemmas. Development of cleaner technologies, mainly of cleaner fuels that would provide a similar energetic content as those used today, is a long-term solution. It is also expected that this solution would release countries from the economic stress of energy security.

In addition, the need to assign sustainability attributes towards methods of producing fuels and how these are transformed into end-use power sources has become an inherent requirement for society. This need has become a point of interest, not only in terms of technical and feasibility concerns, but because knowing such attributes will enable scientists and policy-makers to reach an in-depth understanding of the benefits of acting proactively, rather than facing the consequences of not doing it so.

Since the mid-1980s, hydrogen was envisioned by researchers and government authorities as the main energy carrier of the future, being used mainly for fertilizer production at the time. In this vision, hydrogen would not only satisfy transportation energy demands, but also become the leading national energy source from renewable origins to power the sought clean economic development (Mattos 1984).

However, a revolutionary vision cannot be laid into policy if economics fails to make hydrogen technologies go in the right direction. These viability questions have emerged in previous research studies that were aimed at analyzing the technological feasibility of hydrogen as an energy carrier and how this will become the foundation of emergent economic structures in future societies (Balat 2009).

Many technologies are available for a commercial production of hydrogen, although such technologies rely on heavy energy input, rare materials for catalytic purposes, and/or high cost of complex manufactured materials. The main impairment on these technologies appears to rely on high costs, market placement, and energy input based on fossil fuels. Other renewable sources such as solar, hydropower, and biomass could shift the heavy energy burden, in order for hydrogen production to be economically and environmentally sound.

Some studies have analyzed the solar hydrogen energy system transition, where hydrogen would be produced mainly from the renewable energy powered by water spillage through electrolysis (Momirlan 2002). However, hydrogen storage and transportation, due to its low volume of energy density, still poses a challenge for large-scale distribution systems that aim to operate efficiently and at a low cost.

Hydrogen is indeed based on its abundant rate, caloric/thermodynamic value, and energy carrier capacity, a fuel that is sought to be harnessed for powering economies of the future. Its current limitations are characterized worldwide by technological drawbacks due its relative new state of development. However, economic support from various countries around the globe have been targeting R&D (Research and Development) efforts for some time, and are already in position to polish and streamline manufacturing technologies/techniques and new alternatives. Although hydrogen production is still on its developing stages, it is clear that its inherent use as a fuel will go beyond the vehicular stage.

The main question to answer throughout this discussion is: ***Does hydrogen production provides a successful framework as an advanced and sustainable fuel***?

This study is going to analyze the complex interaction between local and international factors in Brazil that drive current renewable fuel demand, focusing on hydrogen. Social issues will take a fundamental part on the analysis, trying to shed light on stakeholder interests and how these are included or dismissed for decision and policy-making. Finally, environmental issues and concerns will cover the third dimension of sustainability for evaluating hydrogen as a sustainable fuel.

By addressing dynamics and everyday changing facts (energy demands, types of energy exploited, natural resource extraction rates, processing of resources, international trade, and market fluctuations) with a thinking analysis system and a wholistic approach, some multi-variable and complex problems are going to be able to find an integrated solution that changes over time, but also adapts to provide an acceptable result at a determined place and time.

6.2 Aims and Objectives

The main aim of this subchapter is to provide a qualitative sustainability measure of current hydrogen production techniques in Brazil. The study is focused on finding adequate social, economic, and environmental indicators to measure such technologies in a qualitative manner. Although some figures are going to be used to account for factors such as GHG (Green House Gas) emissions, the analysis will be focused on the strength of the indicators and how much weight they can place within a decision-making process. To accomplish the analysis, indicators will be weighed against possible options for hydrogen-producing technologies through a multi-criteria analysis. The results are intended to streamline the right indicators, providing valuable stakeholder information for decision-making purposes within public or private fields.

In order to achieve the outcome of the multi-criteria analysis, two subordinate objectives must be previously accomplished:

1. *Establishment of a benchmark that will serve as a reference point for the analysis. The benchmark should symbolize what hydrogen might accomplish by its replacement for existing fuel sources.*
2. *Formulation of a decision-making framework where hydrogen production options will be proposed. The options will represent hydrogen production methods in Brazil and are to be assessed through the MCA (multi-criteria analysis) methodology. Analysis results should highlight their sustainability features and also point out which technology or approach is the most promising.*

6.3 Methodology

The following subchapter outlines the selected methodology to be followed for the proposed work. The methodology is aimed at extracting the necessary information to fulfill the requirements setup by the aims and objectives. The established steps were derived from existing MCA literature and by further analysis of how the MCA methodology has been applied to different situations, laying emphasis on the type of information intended to be obtained and on how it was extracted. The outcome of the methodology analysis yielded the set of steps depicted in Fig. 6.1, being adjusted to the particular focus of this work.

Fig. 6.1 Established methodology for the proposed work

6.3.1 Multi-criteria Analysis (MCA) Theory

In order to address the complexity arising from cost and issues of relevance such as environmental impacts which cannot be accurately assessed due to the inequality of their units, as well as their nature and difficulty in establishing physical limits, the multi-criteria analysis (MCA) can assist in providing a sound understanding of the variables and stakeholders at hand. By doing so, it affords the opportunity of a detailed analysis in a better-suited framework where information appears in a structured manner and an equitable unbiased evaluation is feasible for decision-making purposes.

The multi-criteria analysis is not a method intended to standardize all variables; instead, it supplies an unrefined view on the different dimensions and multiple effects of a particular interest (policy, project, investment, direction). Although it can integrate monetary aspects into a determined assessment, the main purpose of its methodology is to provide an integrated understanding of a process instead of a mere economic or cost-benefit evaluation (Hirschfeld et al. 2011).

The main advantage of using the MCA is its capability of combining cost, benefits, and positive and negative aspects of different options where multiple conflicting criteria such as environmental, economic, and/or social issues can be incorporated into the same analysis. The criteria can be then measured if deemed appropriate, and consequently weighted in a performance matrix (Gamper et al. 2006).

6.3.2 Approach

A full literature review is going to be conducted so as to provide a cross-reference analysis of existing sustainability indicators. Emphasis will be given to how indicators have been used to make an evaluation analysis within different frameworks and how to obtain different streams of information.

The first step is to analyze the existing information on current technologies being utilized in Brazil for hydrogen production, regardless of whether such technologies are in operation or in their development state. Current energy policy measures and previous actions towards the introduction of renewable fuels will serve as supporting tools to evaluate stakeholder input along the analysis. A first framework will be obtained at this point, where sustainability indicators are going to be narrowed down to fit the particular characteristics of the analysis, leading the way to establish a preliminary criteria matrix.

The second step is to include information analysis on the website (Guaratinguetá, Brazil) from different local or international sources in order to cross-reference the reviewed information and perform a complete MCA. Identified sustainability criteria and indicators will be applied towards hydrogen production techniques from selected options based on the Brazilian current energy needs.

The information obtained from the multi-criteria analysis is expected to provide a better understanding of hydrogen production from different sources in terms of sustainability. It is important to note that the results obtained will not resemble those of a life cycle analysis (LCA) where measurements are usually quantitative and account mainly for harmful emissions and negative environmental impacts. In this study, the use of MCA as a tool will lean the analysis towards performing a qualitative measurement of sustainability indicators surrounding the decision-making process of implementation, and scale up hydrogen production techniques by means of policy and other social components. It is also intended to be simple enough for policy- or decision-makers, so that it can be used as a whole or in parts by selecting indicators for evaluation as needed.

Thirdly, a summary of the findings is going to be provided along with a critical analysis and possible scenarios for integrating results into the social or economic structure of Brazil. The main purpose of such integration is to provide both the industrial and government sectors with a clear path to understand sustainability features from hydrogen production technologies.

Some studies have used the multi-actor multi-criteria analysis for biofuel applications, and its further integration into demanding markets were triggered by policy and regulation (Turcksin et al. 2011). Although such analysis yielded requirements for a successful integration of renewable fuels into the targeted social schemes in the near future, the purpose of this analysis is to obtain qualitative information on whether hydrogen production techniques could be a sustainable option for Brazil.

In order to achieve an appropriate assessment, the proposed methodology starts with the formulation of a decision framework where options representing hydrogen production techniques are identified. The options are to be assessed and weighted through the use of selected sustainability indicators. Sustainability indicators are to be screened, selected, and, if required, enhanced from existing indicators representing the three main pillars of sustainability. The screening is to be based on criteria fitting the proposed framework, and the critical literature review on policy, decision-making, and renewable fuels. The following methodology (Fig. 6.1) is an extract from (Gamper et al. 2006) with additional points in order to fit the framework of the proposed work.

6.3.3 Limitations

The limitations of the present work comprise the reduced amount of empirical data for current and new hydrogen production technologies, which would increase the quality of the results. Although there is a substantial amount of studies performed by countries around the world regarding hydrogen production, applications for its use, transport, etc., most of them have only focused on research or pilot levels and have not leaped on an industrial scale. Stakeholder involvement is a valuable asset for this work and, if possible, interviews will be conducted for data collection purposes, however, the time and resources for such project might also limit the results obtainment.

The system boundaries for the analysis extend from basic components of fuels to their corresponding energy and material streams for extraction, processing, refining, storage, transportation, sale, and end use by the consumer. The former inclusions are necessary for a complete and integrated analysis. Although information might not be available, educated guesses will be made to provide variables with a value.

6.4 Background on Renewable Energy and Hydrogen Production in Brazil

The following subchapter discusses historic and current uses of renewable energy sources in Brazil, focusing on hydrogen production techniques. The ethanol industry is explained exhaustively with the purpose of unveiling key moments in history of policy-making towards this renewable energy source, and how these efforts were able to establish ethanol as a primary fuel for quite some time. Understanding the uprising of sugarcane and ethanol industries in Brazil becomes of great importance when other renewable sources of energy or fuel are considered for integration or substitution in the consumer market by means of policy-making.

6.4.1 Introduction and Use of Renewable Fuels in Brazil

Brazil could be considered a pioneer as regards the use of biomass-based renewable fuels, as they have been consumed since the beginning of the twentieth century. As sugarcane production and harvesting were already an established trade for manufacturing sugar, the use of ethanol as a fuel was prioritized as a measure so as to liberate Brazil from its dependency on imported paraffin, an issue that became increasingly outstanding to the point of labeling it as "the national fuel" by the state of Pernambuco by the year 1919 (Galli 2011).

Brazil could be regarded as a privileged country due to possessing the second largest hydropower potential in the globe. This advantage played an important role during the first oil crisis where hydropower participation in total energy consumption rose from 19 % in 1973 to 29 % in 1983. Such an increase aided the country in substituting fossil fuel resources for electricity generation purposes (Mattos 1984). Hydropower currently represents the main source of electricity in Brazil, which has been displacing the use of fossil fuels for electric generation purposes (Fig. 6.2). Hydropower, considered by a vast majority as a renewable source of energy, is identified as a viable candidate for powering other manufacturing processes of first, second, and third generation biofuels. The estimated hydropower potential in Brazil is around 250,000 MW, however only 30 % of this potential has been used due to policy restrictions that protect land conservation units and reserves for indigenous populations, where the largest hydropower potential is concentrated within the Amazon River basin (Brazil Works 2012).

Brazil's energy policy is currently aimed at supporting expanding hydropower capacity, oil exploration, and extraction of newly found reserves, as well as ongoing expansion of biofuel (ethanol, biodiesel) production and national energy efficiency measures. Brazil is set to become the largest exporter of ethanol in the world.

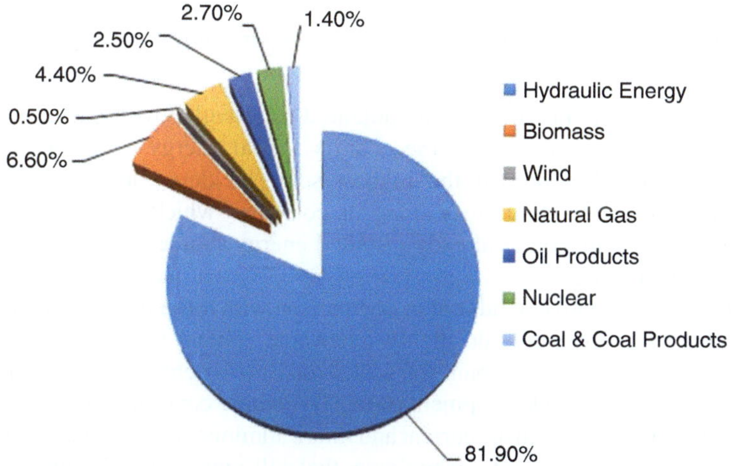

Fig. 6.2 Brazil electric energy offer by source, 2011 (Reproduced from MME 2012)

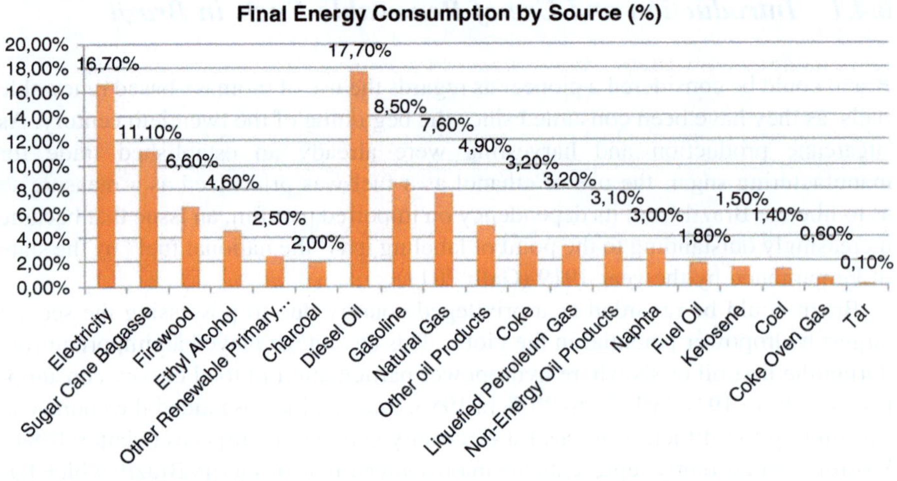

Fig. 6.3 Final energy consumption by source, 2011 (Reproduced from MME 2012)

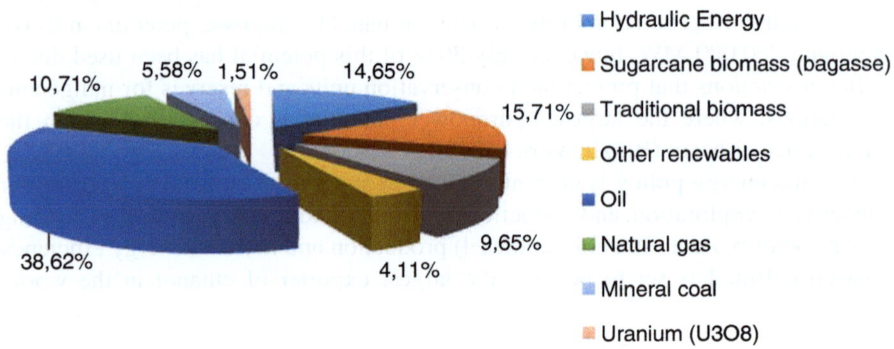

Fig. 6.4 Brazil energy matrix, 2011 (Reproduced from MME 2012)

However, its renewable generation potential is greatly overlooked within the energy policy and confirmed by the country's final energy consumption matrix (Fig. 6.3). Brazil has one of the highest solar incidence areas in the world, accompanied by wide wind areas along its coastline, which have been proved to be competitive against other already installed energy sources (Brazil Works 2012; International Rivers 2012).

Brazil's energy matrix stands out in comparison with those from highly developed nations due to its diversity and highly renewable content (Fig. 6.4). As in 2011, renewable energy sources account for 44.1 % of Brazil's energy matrix, while Economic Cooperation and Development (OECD) member countries have only reached 8 % (USDA 2012). It is up to the current and future administrations to make appropriate shifts in policy to accommodate technologies that will continue to drive the country in a positive direction with regard to renewable energy generation and use.

After the first oil crisis in 1973 where the cost for imported oil increased from \$2.7 USD/barrel to \$11.70 USD/barrel, Brazil's foreign debt was severely impacted, affecting not only the balance of trade, but also provoking high inflation during the following years. In response to evident high oil prices and the threat of economic instability, the Brazilian government launched three major projects: (1) national oil exploration and production; (2) large-scale expansion of hydroelectricity generation; and (3) development of substitutes for the three major oil sub-products: diesel, fuel oil, and gasoline (Cerqueira Leite et al. 2009).

The Proalcohol program, one of the national measures taken in 1975, aimed to slow down energy consumption by means of ethanol production from biomass sources. It succeeded to prove its large-scale ethanol production from sugarcane and its further use as a substitute for gasoline in combustion engine vehicles (Lèbre and Santos 2011).

The program was deployed in two phases, the first one started by selecting sugarcane as the main feedstock for ethanol production followed by setting a fuel standard to mix up to 22.4 % (by volume) anhydrous ethanol on all gasoline sold in the country. Phase 2 was characterized by supporting initial measures for fuel mix through government subsidies that targeted increasing production and distribution of ethanol (Soccol et al. 2005). This phase was marked by an increased expansion of sugarcane mills and distilleries, and it was reinforced by the capability of sugar mills to produce sugar or ethanol depending on demand and market price, while anhydrous ethanol mix ratios were still flexible in terms of car efficiency. Furthermore, agreements with car manufacturing companies boosted ethanol-only cars, which reached 94.4 % of total automobile production in 1986 (Lèbre et al. 2011).

After 1986, other phases which did not pertain directly to the Proalcohol program developed within the ethanol and car manufacture industries. Phase 3 (after 1986) was marked by a decrease in ethanol production, followed by a major ethanol supply crisis that deteriorated trust in the consumer market as regards ethanol as the main fuel for vehicular use. As a consequence, the ethanol fuel car share plummeted to 1.02 %. Phase 4, from 1989 to 2003 was characterized by standardization in ethanol fuel mixing (up to 24 %) and awareness of environmental benefits of using ethanol as a fuel additive. After 1999, the market price of ethanol has been the main driver for production and demand efforts.

Phase 5 (after 2003) encompassed the need for ethanol as a renewable fuel in the mist of high oil prices, energy insecurity, an established ethanol production infrastructure that could shift current paradigms and the highest flex fuel vehicle fleet, thus creating the required local demand for a circular economy. International concerns for climate change stimulate global ethanol demand and pose a great opportunity for Brazil as the second largest producer and potential largest exporter (Lèbre and Santos 2011).

Due to the national constraints and pressure from international markets on ethanol and oil, the Brazilian government has targeted energy security and economic stability as the core of national energy policies. This trend has been visible since the establishment of the Proalcohol program, and recently on Brazil's federal government support and financing on hydrogen programs since the early 2000s.

In Brazil, as well as around the world, the main uses of hydrogen comprise ammonia production (55 %), oil products refining (25 %), methanol production, and other uses (20 %), totaling 51 million tons of hydrogen per year (CCC 2010). Hydrogen fuel cells are one of the main research and development targets for hydrogen use as a fuel, mainly due to its energy efficiency (between 40 and 60 %) and zero-emission factor. The fuel cell application has quickly spread as pilot programs in densely populated areas, where hydrogen fuel cell powered busses are already in operation. However, other options are also underway such as a direct use of available alcohols in fuel cells (methanol and ethanol), which could, in turn, resolve some of the technical issues imposed by current hydrogen storage and transportation systems.

Direct use of ethanol in direct ethanol fuel cells (DEFC) overcomes storage and infrastructure obstacles placed by hydrogen transformation from other biomass sources. DEFC present several advantages over the already existing direct methanol fuel cell (DMFC), displacing the toxicity properties of methanol, higher energy density 8.0 vs. 6.1 kWh/kg for ethanol and methanol, respectively, and higher CO_2 sequestration from root microorganisms of sugarcane harvesting (Hotza et al. 2008).

6.4.2 Hydrogen

Although hydrogen is not a widespread used fuel for vehicles and industrial power generation purposes, its presence has been on the rise not only in Brazil, but also internationally. Most activities in Brazil since the late 1980s have been focused on research, but it was not until 2002 when the federal government started a Fuel Cell Program. The program ($ProH_2$),[1] supported mainly by the Ministry of Mines and Energy and the Ministry of Science and Technology, aimed to make Brazil internationally competitive by supporting cooperative research and development for fuel cell production and hydrogen storage.

Hydrogen and fuel cell systems provide a large flexibility as fuel sources based on the available technologies for conversion and processing. Given the large amount of renewable energy resources available in Brazil, hydrogen production based on such renewables allows an apparent sustainable conversion from biomass. In regions where renewable energy resources are large, hydrogen can be produced and stored for further transport to low-energy resource areas such as large regional centers, where it would serve as transportation fuel or for energy generation purposes (Hotza 2008).

Conversion from chemical energy into work comes into consideration when evaluations of energy efficiency are required. This is particularly valid in the case of propulsion systems for vehicles. In the case of hydrogen, fuel cells have been

[1] Formerly "Procac", but renamed $ProH_2$ in 2005 (Hotza 2008).

selected as the main propulsion system in road vehicles due to their stack modular capability for storage purposes and reduced space required for system installation. Studies have found an energy efficiency range for fuel cells of 0.4–0.6 in contrast to internal combustion engines (ICE), where efficiency ranges lay within 0.2 and 0.3 (Granovskii and Dincer 2006). Commercial hydrogen can be obtained from different avenues depending on the material and energy sources being utilized. Based on a technological approach, hydrogen production can be classified in electrochemical, photo-biological, photo-electrochemical, and thermochemical generation.

6.4.3 Steam Reforming of Natural Gas for Hydrogen Production

Currently the main industrial avenue to produce hydrogen in an economical fashion is steam reforming of natural gas.[1] The reaction occurs at high temperatures (700–1000 °C), where steam reacts with methane to produce carbon monoxide and hydrogen gas (Fig. 6.5), according to the following reactions (Gaudernack 1998).

$$CH_4 + H_2O \rightarrow CO + 3H_2 \tag{6.1}$$

$$CO + H_2O \rightarrow CO_2 + H_2 \tag{6.2}$$

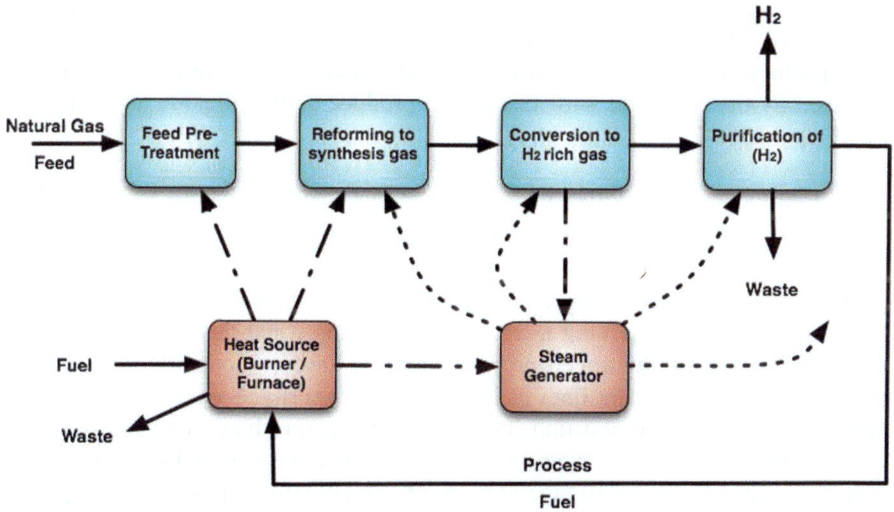

Fig. 6.5 Steam reforming of natural gas for hydrogen production flowchart (Modified from Molburg and Doctor 2003)

For the overall reaction:

$$CH_4 + 2H_2O \rightarrow CO_2 + 4H_2 \quad (6.3)$$

Partial oxidation of methane (CH_4) is also an intermediate process for hydrogen production, where the proportion of hydrogen to hydrocarbon is greater to that of the steam reforming reaction.

$$CH_4 + ½O_2 \rightarrow CO + 2H_2 \quad (6.4)$$

Since steam reforming is highly endothermic and partial oxidation is exothermic, combined processes will be suited to achieve higher efficiencies on total production.

As in 2012, the steam reforming of natural gas accounts for 50 % of world's total hydrogen production (Palma et al. 2012).

6.4.4 Steam Reforming of Ethanol for Hydrogen Production

An alternative hydrogen production method based on large hydrocarbons has been suggested by several studies. The case of ethanol has been widely used due to its great abundance in the Brazilian market, while at the same time being an emergent renewable and low cost fuel that will most likely spread and penetrate the European and Asian markets.

Hydrogen production based on the steam reforming of ethanol is similar to that of natural gas steam reforming. The process is characterized by the reaction of superheated ethanol with steam at high temperatures (600–700 °C as an optimum range) where a rupture of the carbon bond occurs, thus yielding CO and H_2, followed by the water-gas shift reaction to produce carbon dioxide and hydrogen gas (Hotza 2008).

$$C_2H_5OH + 3H_2O \rightarrow 2CO + 6H_2 \quad (6.1)$$

$$CO + H_2O \rightarrow CO_2 + H_2 \quad (6.2)$$

As in the steam reforming of natural gas, carbon monoxide can emerge as a by-product other than serving as reactant during the water-gas shift reaction and further persist in the hydrogen product streams. Carbon monoxide may exist in traces in hydrogen product lines as an undesired impurity. Some end use applications such as internal combustion engines, burners, and turbines are not affected by such an impurity. However hydrogen fuel cells which have started to gain momentum in transportation and storage applications (especially in proton exchange fuel cells) can be severely impacted by hydrogen sources with carbon monoxide impurities targeting specialized polymers membranes or catalytic materials.

The utilization of hydrogen for electric power generation in a proton exchange membrane fuel cell (PEMFC) requires the anode inlet H_2 gas stream to contain a CO concentration lower than 10 μmol/mol. Carbon monoxide acts as a poison to the fuel cell platinum electrocatalyst (Sordi et al. 2009).

6.4.5 Hydrogen Production by Electrolysis

Within an electrochemical classification, the most utilized industrial process for hydrogen production today is water electrolysis. Hydrogen is produced through water electrolysis by splitting water molecules into hydrogen (H_2) and oxygen (O_2) as depicted in Fig. 6.6. The process takes places within an electrolytic cell where two partial reactions occur at two separate electrodes. The electrodes are submerged into an ion-conducting electrolyte where hydrogen is produced at the negative electrode (anode) and oxygen at the positive electrode (cathode). The required charge exchange to split water molecules occurs through the flow of OH-ions (aqueous KOH saline electrolyte solution) and electric current within the circuit (Silveira et al. 2009).

There are high energy requirements for electrolysis in the form of electric power, therefore high production rates of hydrogen can become economically unfeasible due to the elevated cost of electricity based on fossil fuels, such as coal or diesel to generate such power. However, the alternative of powering massive electrolysis arrangements with a combination of renewable energy sources such as solar and wind is possibly an economical alternative for hydrogen production (Turner 2004).

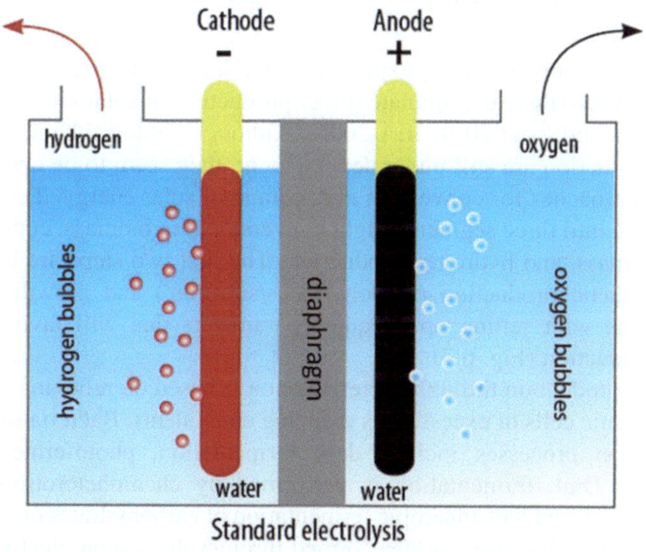

Fig. 6.6 Water electrolysis for hydrogen production (Reproduced from Hydroxsystems 2013)

6.4.6 Hydrogen Production by Pyrolysis/Gasification

Pyrolysis refers to the thermochemical breakdown of complex hydrocarbons or biomass at high temperatures in the absence of oxygen. Decomposition of organic matter through this process yields liquid and gas products and a residue which is rich in carbon, such as ash or tar. The liquid product termed "*biocrude*" is a mixture of aldehydes, alcohols, acids, and oligomers from the original carbohydrates and lignin biomass along with water from dehydration reactions. Hydrogen can be then obtained by steam reforming the biocrude (Mann et al. n.d.).

Gasification refers to the transformation of biomass or fossil-based hydrocarbons into carbon monoxide, hydrogen, and carbon dioxide. The process takes places at elevated temperatures (above 700 °C) with a controlled amount of oxygen and without promoting combustion. The partial oxidation of the components yields a gas mixture called syngas (synthesis gas), which can be then steam-reformed into hydrogen (FCHEA n.d.).

Pyrolysis or gasification of biomass presents a particular advantage in Brazil since most of the dry weight of crushed sugarcane (bagasse) is used as burning fuel for cogeneration purposes in sugar mills and ethanol refineries. Using bagasse as a feedstock for pyrolysis will increase hydrogen production, but it will deprive sugar mills and ethanol refineries from an already established source of biomass energy, which currently aids the ethanol economy to lower CO_2 emissions from fossil fuel use.

6.4.7 Hydrogen Production by Biological Processes

New technologies are also being explored, which include the use of photosynthetic bacteria and macroalgae to stimulate direct production of solar energy into hydrogen (Srirangan and Pyne 2011; IEA 2005). Although photosynthetic processes for hydrogen production are still under development, they seem to be one of the most promising approaches for conversion and storage of solar energy. The mechanism can be divided into three segments, light conversion into biomass, concentration of substrate/biomass, and hydrogen production. The first two steps are characterized by photosynthetic production (carbohydrates/substrate) and growth of algae or bacteria, along with setting up adequate parameters that will favor an optimal hydrogen production (Fig. 6.7).

Hydrogen production through bacteria action is based on relieving fermentative or photosynthetic cells of excess, thus reducing equivalents. Bacteria-driven hydrogen production processes include dark fermentation, photofermentation, and biophotolysis. Dark fermentation is performed by chemoheterotrophic species where a light-independent anaerobic fermentation of carbohydrates or other organic substrates occurs. Hydrogen is then formed through the proton-electron exchange from substrate catabolism. Hydrogen can also be produced through

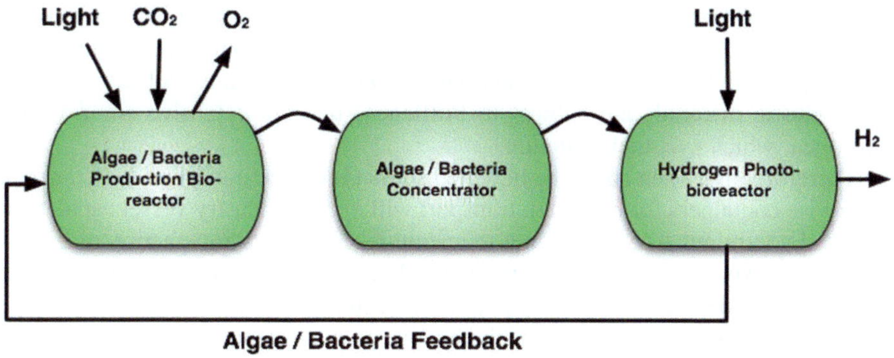

Fig. 6.7 Representation of biological hydrogen production (Reproduced from IEA 2005)

photofermenting where non-oxygenic bacteria use alternate reducing compounds, such as organic acids and hydrogen sulfides as electron donors. Purple non-sulfur and green sulfur bacteria are involved to photo-heterotrophically convert substrate compounds into hydrogen and CO_2 by means of solar energy.

Biophotolysis is exclusive of algae and cyanobacteria species that have photo-autotrophic capabilities. Such species show the highest rates of hydrogen production from solar energy, and due to their potential application to small or nonarable land areas, are being subject of great interest for countries that have a high incidence of solar radiation. Solar radiation is required in biophotolysis for oxygenic photosynthesis where water acts as the electron donor. Through water splitting, protons are then generated and further reduced to molecular hydrogen (Srirangan and Pyne 2011).

6.4.8 Hydrogen Storage and Distribution

Hydrogen has the highest energy density by weight, but a low energy density by volume. Being the lightest element due to its low energy density, the transport of large amounts of energy in the form of gaseous hydrogen can become challenging and accrue an elevated cost.[2] Pipelines, pressurized cylinders, and tankers are currently the main transport means for hydrogen.

Pipelines for hydrogen transport have been in existence for over 50 years, both in the USA and Europe. The existing pipeline is mostly used to transport hydrogen from the source to manufacturing facilities where the product is required as a raw material. Hydrogen pipelines usually operate at a pressure of 10–20 bar, but can also operate at pressures of up to 100 bar. Hydrogen pipelines must be

[2] Energy required to transport hydrogen through a pipeline is on average 4.6 times higher per unit of energy than that for natural gas (IEA 2005).

manufactured from nonporous materials, such as stainless steel, so as to avoid permeation and eventual loss of hydrogen. Materials used to transport natural gas, such as polyethylene, are not suitable for hydrogen transportation. Polyethylene and other plastic materials are subject to brittleness through extensive contact with hydrogen, which eventually rupture.

Therefore, hydrogen has to be liquefied so that it can be transported via tankers, otherwise containers for hydrogen transport will exceed industry vessel capacity at standard conditions of pressure and temperature. However, liquefaction of hydrogen is an expensive process, particularly because hydrogen needs to achieve cryogenic temperatures (−253 °C). Energy requirements to cool and pressurize hydrogen can reach up to 31 % of the energy content of liquid hydrogen, with a best-case scenario of 21 % when the electric power for liquefaction was generated with 50 % efficiency (IEA 2005).

6.5 Multi-criteria Analysis

The following section discusses the methodology approach described in Sect. 6.3.2 in detail. Emphasis is given into fitting multi-criteria analysis methodology into a particular decision-making context.

6.5.1 Establishing a Decision-Making Context

On 1987, the World Commission on Environment and Development (Brundtland Commission) published its report introducing the definition of sustainable development as "*development that meets the needs of the current generations without compromising the ability of future generations to meet their own needs*" (WCAD 1987), a strong and encompassing phrase that has led the way through environmental regulations, policy-making, and changes in lifestyle. However, the sustainable development concept although already coined requires to be molded in order to fit particular systems or decision-making context situations.

Hence, by taking the root of what sustainable development portrays, it can be found the right set of characteristics that will define the present particular point of analysis. While information about sustainability and sustainable development is now plentiful, the definition of "sustainable fuel" or what "sustainable fuel" is supposed to represent is yet to be discovered indirectly between the lines of scientific journals. Perhaps the reason is not because characteristics to define the term are lacking in, but mostly on account of the various implications of elaborating a representation that would include all the desired requirements, without compromising interests of economic or political parties.

The previous statement gives rise to an issue, since without a perfectly clear definition of what "sustainable fuel" entails, it becomes difficult to achieve local or international consensus; hence further cooperation and commitment agreements will be deemed challenging, if not impossible. For the purpose of this work, a definition of "sustainable fuel" will be established. It is not aimed at imposing a universal brand, but it rather serves as a benchmark for analysis purposes on the present study. This benchmark will represent a convergence point for the multi-criteria analysis options and will assist in narrowing down indicators and criteria that, despite possessing a general importance, are not essential for the present analysis.

Thus, sustainable fuel is defined as a liquid or gaseous fuel which has been generated from at least 80 % local renewable energy resources and is utilized mainly to power vehicular and mechanical machinery of internal combustion engine or fuel cell technology nature. The end use of such fuel shall reduce overall negative environmental impacts in at least 60 %, if compared to using such fuels from nonrenewable sources.

Once a benchmark is set, it is possible to start an analysis based on existing technologies for hydrogen production. It is important to note that the focus will be to push hydrogen as a leading fuel. Existing or technologies under development will be assessed within the Brazilian economic, technological, and social frameworks and the reason why this should or should not be the fuel that ought to power the Brazilian economy in the future is going to be evaluated.

By performing an analysis, questions will be raised as regards important national and international factors. Although most factors will be accounted for, focus will be given on the following points, as they have been identified as main drivers for shifting current "business as usual" schemes towards sustainable fuel production and use:

- Provide Brazil with energy security, improving resilience and preventing the national economy from being affected through a carbon bubble originating from an international overvaluation of estimated petroleum reserves and forecasted extraction flows (The Guardian 2013a, b, c).
- Political pressure to comply with international emission agreements to reduce GHG without hurting Brazilian economic and social structures.
- Environmental concerns to improve air quality in largest metropolitan areas in Brazil where health quality is compromised by substantial amounts of toxic compounds lingering in the air as a result of automobile exhausts.
- Revamping power generation techniques that avoid the use of fossil fuels and focus on renewable sources such as biomass, wind, solar, and tidal (McLellan et al. 2005).
- Evaluation of current fuel storage and transport infrastructures and assessment of whether new fuels will be suitably integrated into existing systems or new ones will have to be developed.
- The impelling need to satisfy a constantly growing fuel demand that will, in time, require oil-derived product imports if no further investments are made in ethanol processing and refining infrastructure or other fuel sources (The Bioenergy Site 2013).

The previous drivers consolidate the decision-making context that is sought to steer the multi-criteria analysis towards the fulfillment of the aims and objectives established in Sect. 6.2.

6.5.2 Identification and Selection of Options

In order to identify a viable model for sustainable fuels, several options need to be assessed providing the sought information by means of comparison and evaluation. Such options are required to portray a standard operation model that, if not already implemented, could be adapted or integrated in the future. The selected options can be characterized as being fossil-fuel based, renewable-energy based, or a combination of both. This hint of complexity is expected to provide flexibility as well as limitations, but mainly a closer resemblance to reality.

The following questions were fundamental to assemble an adequate set of options for the analysis. The questions originate from the concept of directing policy or other social tools to achieve a "sustainable fuel" status, hence providing a way to obtain key information that will assist in doing so. The selected set of options should be able to partially answer these questions, as well as to expand the required knowledge to reduce uncertainty factors.

The São Paulo metropolitan region is categorized as a major source of air pollution, a result of overabundance of diesel buses and light vehicles crowding streets to mobilize 18 million residents. Average concentrations from 1997 to 2003 reached 49.0 $\mu g/m^3$ for PM_{10} and 43.1 ppb for ozone (Bell et al. 2006), where correlations have also been found between PM and bronchitis amongst other illness (Ribeiro and Cardoso 2003), attributes that are a direct effect of the transportation sector. Federal authorities as well as international actors are involved in implementing the Environmental Energy Strategy: Buses with Hydrogen Fuel Cell Project (Silveira et al. 2009).

- How is characterizing a fuel as being sustainable going to enhance society's performance?
- How will societies set standards, track changes, and improvements through policy-making?
- How can policy-making encourage international dialogue, expertise exchange, and identify critical thresholds?

The previous questions will also act as system boundaries for the analysis, as it is sought that identified indicators will couple towards the selected options to obtain the required answers.

6.5.2.1 Hydrogen from Coal Gasification with Carbon Capture (HCGCC)/Option 1

This option represents a centralized production scheme, where hydrogen is obtained through gasification of coal followed by removal of CO_2 from the product stream. The captured carbon will be then stored in an approved reservoir. The uncertainties for this option are nearly absolute, as for the selection and leak proof reliability of carbon reservoirs for injecting CO_2. Literature indicates that there is still research and development yet to be fostered, as the apparent cost of sequestering a ton of CO_2 from gasification processes remains high (FCHEA n.d.). Hydrogen transport and storage will have to be developed or assessed if current distribution systems are adequate to accept hydrogen gas at standard conditions of pressure and temperature.

6.5.2.2 Hydrogen from Electrolysis Powered by Renewable Sources (HEPRS)/Option 2

The option represents both a centralized and decentralized scheme, where economic electrolysis units are developed for hydrogen production within industrial sites or rural areas. Power supply for electrolysis units is to be provided by the most abundant local renewable energy source (sugarcane bagasse, photovoltaic cells, wind turbines, or a combination of any). Methane from anaerobic biodigesters or landfills is not taken into account, since methane can also be used as a substitute for fossil fuels. Storage and distribution systems are to be developed if a decentralized scheme is adopted, however the cost is assumed to remain low, as it will be pro-rated by eventual growth in infrastructure. If a centralized scheme were adopted, current distribution and storage networks should be evaluated for compatibility.

6.5.2.3 Hydrogen from Biological Processes [Biophotolysis] (HBP)/ Option 3

The option represents a centralized scheme where vast farms of photoautotrophic macroalgae and/or cyanobacteria are maintained at optimum conditions for hydrogen harvesting. The centralized scheme is necessary for both algae and bacteria strains utilized are solely photoautotrophic, hence the need to isolate hydrogen production in highly solar irradiated zones. It is also limited by close-by intensive CO_2 sources, as some algae strains require an external carbon source to maintain metabolic growth. The scheme is challenged by investments in new distribution and storage infrastructure, as no existing infrastructure is available. The time frame for this option to be operational is also a limiting factor.

6.5.2.4 Hydrogen from the Steam Reforming of Natural Gas (HSRNG)/Option 4

The option represents current procedures for hydrogen production based on fossil fuel sources and no carbon capture implementation. A business-as-usual scenario will serve as a contrast point within the MCA weighting matrix. Natural gas is assumed to come from national origin and transported directly after the steam reforming process. Current storage and distribution systems are used to transport natural gas and an assessment is required to use the same system for hydrogen distribution. No investments are required for this scheme, nor the need to reach a determined timeframe.

6.5.2.5 Hydrogen from the Steam Reforming of Ethanol (HSRE)/Option 5

This option represents the use of current steam reforming of natural gas process with feedstock substitution for hydrated or anhydrous ethanol from sugarcane. For this scheme, it will be assumed that there is an overage in ethanol production that is not going to disturb current demands for vehicular ethanol nationwide. This option may represent significant advantages since a full infrastructure to support ethanol production already exists, as well as alternate demands for use and marketing of ethanol which assists in stabilizing the price of ethanol within the national market. A centralized or decentralized scheme can be adopted for ethanol distilleries that are scattered throughout sugarcane harvesting areas or that can also be largely found in metropolitan areas.

6.5.3 Criteria for Indicator Selection

Discussing sustainability indicators can lead to subjective thinking, inaccuracy, and lack of reliability. Underlying reasons arise from the fact that some indicators proposed in previous studies are not easily measured, understood, or are simply not a proper fit for the study framework at hand. In order to select the right set of sustainability indicators for this analysis, a set of screening criteria was used as follows.

Relevance was pinpointed as one of the main criteria, as some indicators did not provide any input towards establishing a value for the options. *Practicality*, as it is desired that qualitative or quantitative data is available for the assessment and that indicators are easy to understand within the framework, especially within the decision-making process. *Technically and scientifically sound*, although indicators might not provide a quantitative measure, they should sustain firm scientific meaning and prove to be technically viable. *Objectivity* will mark the final criteria

Table 6.1 List of overarching indicators

Environmental	Economic	Social
1. Potential to reduce GHG and toxic emissions	5. Provide energy security and economic resilience towards international market fluctuations	9. Potential to improve health
2. Potential to substitute fuel sources of electricity generating technologies	6. Adaptability of current storage and distribution infrastructure	10. Potential to increase quality of life
3. Land use impact and land use change	7. Adaptability for industrial and vehicular use	11. Potential to improve income levels
4. Material and energy intensity	8. Required investment for infrastructure	12. Increase of trade specific jobs

for indicator selection; as for validity and quality of multi-criteria analysis results to be obtained, an unbiased assessment of the options is required.

The main objective in assembling a set of quality indicators is to provide an accurate depiction and expose the true potential of each proposed option under the present framework. Such information is meant to encourage sustainable production and use of fuels, and also to inform decision-makers on how to modify or create policies that will support these initiatives (Azapagic and Perdan 2000; FAO 2011).

6.5.4 Indicators for Sustainability Assessment

Once criteria are established, it is now possible to identify and select a group of sustainability indicators that better suits the proposed framework and options. A first selection matrix was created with extracts from Azapagic and Perdan (2000) and Carrera and Mack (2010) (Appendix A). The selection process, as well as the merging of indicators, was conducted based on the criteria established in Sect. 5.3, which specifies framework requirements. The main intent of merging indicators is to provide overarching components that strengthen the selected indicators, thus providing a full integration of proposed options into the assessment. Finally, a set of 12 sustainability indicators was selected for the assessment of options (Table 6.1), which represent the three pillars of sustainability fundamentals.

6.5.4.1 Environmental Indicators

- ***Potential to reduce GHG and toxic emissions*** (%)

  It is one of the fundamental indicators for the analysis, as it will provide an overview of greenhouse gas reduction in both production of fuels and their end use. It is important to note that the indicator also accounts for toxic emissions that will directly affect users, even in rural places where decentralized schemes are adopted. Sugarcane burning is not accounted for within the indicator, due to its phasing out from mechanized harvesting of sugarcane.

- ***Potential to substitute fuel sources of electricity generation technologies (1 = very low, 5 = very high)***

  In order to classify a fuel as sustainable, it takes enough chemical and physical flexibility to provide a similar energetic value for electricity generation purposes, as well as adaptability to existing systems and technologies that would need minimal or no adjustment for an acceptable coupling.

- ***Land use impact and land use change (1 = very low, 5 = very high)***

  This indicator was selected in order to assess how fuel production and use impact different land use areas in Brazil, as large areas are required for the production of some fuels, which will keep expanding as local and international demands rise. It becomes important to assess how land use change will impact not only fuel production, but also the environmental and social impacts as a result of such changes.

- ***Material and energy intensity (1 = very low, 5 = very high)***

  This indicator was selected to provide an overview of the raw materials and energy used for production, storage, and distribution of fuels. Both the material and energy sources are assessed based on their origin, previous processing, chemicals, as wells as the water and other resources involved. This indicator is expected to provide a rough figure on the compound material-energy matrix for every liter or gallon of fuel produced and delivered to consumers.

6.5.4.2 Economic Indicators

- ***Provide energy security and economic resilience towards international market fluctuations (1 = very low, 5 = very high)***

  This indicator is based on the proficiency a fuel has to place itself within local and international markets. The outcome is a result of current infrastructures being in place for production and distribution purposes, but also for their availability to be used by the majority of population. The aforementioned is portrayed by the large amount of existing flex fuel vehicles that were currently deployed within the Brazilian market.

- ***Adaptability of current storage and distribution infrastructure (1 = very low, 5 = very high)***

  The indicator addresses potential changes in use for the current distribution infrastructure, such as underground/overground piping, fuel wagon, ship/vessel, or motor methods for distributing fossil fuels among the ones within the proposed options.

- ***Adaptability for industrial and vehicular use (1 = very low, 5 = very high)***

  The present indicator addresses the potential substitution of fossil fuels within different sectors of industry as well as light vehicles. It also evaluates the required changes that current industrial and vehicular technologies will have to undergo in order to accept the new proposed fuel.

- ***Required investment for infrastructure*** **(*1* = *very low*, *5* = *very high*)**
  The aforementioned indicator addresses the need for infrastructure on a monetary fashion depending on new fuel adaptability's potential. The indicator will reflect how intensive the required investment will be in order to launch the proposed fuel.

6.5.4.3 Social Indicators

- ***Potential to improve health*** **(*%*)**
  This social indicator will assess improvements in health based on the potential to reduce admittance of patients suffering from respiratory diseases to hospitals located in densely populated areas. It will be assumed that respiratory sickness is a direct effect of air pollution caused by internal combustion engines' exhaust system. The available data will cover the admittance of both adults and infants.
- ***Potential to increase quality of life*** **(*1* = *very low*, *5* = *very high*)**
  The present indicator will assess improvements on the quality of life, including noise reduction, air quality, cheaper fuels and improvements in machinery or vehicle maintenance time.
- ***Potential to improve income levels*** **(*1* = *very low*, *5* = *very high*)**
  This indicator may prove difficult to be assessed, as the economic success of the proposed fuel is not going to be entirely achieved until a stabilization period within the economy is reached. It will provide an educated guess based on social factors as to how the new proposed fuel will promote adjusted income levels and a better distribution of resources.
- ***Increase on trade-specific jobs*** **(*1* = *very low*, *5* = *very high*)**
  The indicator presented above will identify the amount of jobs created based on the implementation of new techniques for fuel production trends. Figures may include direct and indirect jobs as a result of technological implementations, as well as those resulting from developing infrastructure.

6.5.5 Performance Matrix

On the performance matrix depicted in Table 6.2, a value is assigned for every option depending on the sustainability indicator being measured in Sect. 6.4.1. The value range depends on the set scale established in Sect. 6.4.1. A justification for values assigned to each option is depicted in Table 6.3.

Table 6.2 Performance matrix of options based on selected indicators from Sect. 6.5.4

Criteria/ indicators[a]	HCGCC/ Option 1	HEPRS/ Option 2	HBP/ Option 3	HSRNG/ Option 4	HSRE/ Option 5
1 (%)[c,e,f,g,h,k]	20	80	90	40	70
2[d,e,f,g,h,j]	5	3	3	5	4
3[d,h,l]	1	4[b]	2	1	3
4[c,f,h]	5	2	2	4	3
5[e,f,h,l]	3	5	5	2	4
6[c,d,e,f,h]	4	2	1	3	3
7[c,d,f,g,h,j,l]	1	3	2	3	4
8[c,d,e,f,h]	3	3	5	3	2
9 (%)[f,h,i,k]	20	60	80	30	50
10[d,f,g,i,k]	1	4	4	1	3
11[d,f,g,l]	1	5	3	1	3
12[d,f,h,l]	1	4	3	2	3

[a]List of overarching indicators from Table 6.1
[b]Area affected is assumed to be arable land
[c]Granovskii et al. (2006)
[d]Hotza and Diniz da Costa (2008)
[e]Gaudernack and Lynum (1998)
[f]McLellan et al. (2005)
[g]Carmo de Lima and Veziroğlu (2001)
[h]IEA (2005)
[i]Ribeiro and Cardoso (2003)
[j]Afgan and Carvalho (2002)
[k]Bell et al. (2006)
[l]Galli (2011)

6.5.6 Weighting of Criteria/Indicators

Table 6.4 shows the weight of each indicator based on its range and also on its importance. A scale between 0 and 100 is used to grade each interval or measuring pointer. Furthermore, indicators are catalogued by importance depending on the reach of their scale. For example, indicator (1) is considered to be a high-importance indicator,[3] hence the lowest grade will be "0" and the highest one would be "100." In contrast, indicator (3) will only achieve a grade of "80" since it is considered as being less important than indicator (1). A high grade will be given to the indicator based on the "ideal realistic value" from Appendix B.

[3] High importance indicators are selected based on their weight within each of the three pillars of sustainability expressed in Table 6.1. Indicators 1, 2 and 4 are assumed to have more weight in decision-making, thus a higher possible scale is assigned to them if compared to indicator 3. Economic indicators 6, 7 and 8 are intentionally given lower possible scales so as to avoid economic decision bias while still maintaining an objective weighting and final assessment.

Table 6.3 Justification of options against sustainability indicators

Criteria	Justification
1 (%)	Hydrogen produced from bacteria/algae will achieve higher reduction than electrolysis powered by renewable sources, followed by steam reforming technologies, and finally by coal gasification[a,c,d,e,f,i]
2	Coal-based and steam reforming technologies have the highest potentials for substitution, as their infrastructure is already setup to process their feedstock. High investments and modifications will not be necessary, but minor adjustments to existing processes will be required[b,c,d,e,f,h]
3	Land use impact will be greater for electrolysis powered by renewable sources if the area affected is arable land where solar panels and wind turbines will be deployed. Biological production is considered to be negligible, as the land can be either arid/desert or artificial floating islands. Coal and steam reforming technologies are considered as having a low impact on land use. Hydraulic power is considered as a renewable source[b,f,j]
4	Hydrogen from coal with carbon capture is considered to have the highest intensity due to resources required for coal gasification and carbon capturing. Steam reforming technologies follow the same pattern due to gas and ethanol being used as feedstock. Biological and electrolysis powered by renewable energy production are considered to have the lowest intensity[a,d,f]
5	As regards the Brazilian reserves of natural resources, natural gas will be depleted before coal. Coal being the resource that will stabilize the economic turmoil if international markets crash. The steam reforming of ethanol, renewable-source electrolysis, and biological production will significantly increase economic resilience if compared to fossil resources[c,d,f,j]
6	Based on the current installed storage and distribution infrastructure, no option will promote a full integration. However, since some of the options involve hydrogen production based on fossil fuels, those options will provide a slighter edge for adapting current transport infrastructures, such as piping. Storage infrastructure will have to be developed for all options[a,b,c,d,f]
7	Based on current studies, no option has full adaptation capabilities. However, Brazil has a track record to impose vehicular fuel sources, thus making the vehicular market shift accordingly. Taking this into consideration the potential of each option to adapt industrial and vehicular use will depend on the feedstock source for hydrogen production. Electric cars are also taken into consideration as a preliminary step to fuel cell propulsion.[a,b,d,e,f,h,j]
8	The highest investment will be required for hydrogen production based on biological hydrogen production, as the technology is still under development. Renewable, coal, and steam reforming of natural gas production have their infrastructure already in place, which will require modification, but not necessarily a revamp. Steam reforming of ethanol will require the least changes, while all options share the need for heavy investment in initial storage capabilities[a,b,c,d,f]
9 (%)	Based on potential to emit (PTE) greenhouse gases and toxic compounds of hydrogen being used as fuel for vehicular and industrial purposes, all options represent a significant decrease in emissions. However, the feedstock or method used for hydrogen production will categorize biological and renewable options based on hydrogen with the greatest potential to improve health, followed by the steam reforming of ethanol, steam reforming of natural gas, and coal gasification [d,f,g,i]
10	Based on the combination of factors for quality of life improvement including required space, social benefits such as national economic improvements and potential for education outreach; biological and renewable hydrogen production

(continued)

Table 6.3 (continued)

Criteria	Justification
	shows the greatest potential, followed by the steam reforming of ethanol, steam reforming of natural gas, and coal gasification. Poor working conditions are taken into account for sugarcane harvesting in the steam reforming of ethanol[b,d,e,g,i]
11	Based on economic conditions, available feedstock, current infrastructure for the extraction of fossil fuels, manufacturing of solar panels, wind turbines, fuel cells, and infrastructure for storage and distribution of hydrogen, potential to improve income levels is assumed to be a function of profitable margins from national sales, stable jobs, and resilience of fuel towards international speculation. Disparity on income levels due to national and international monopolization of assets is assumed for coal gasification and steam reforming of natural gas. Renewable-powered electrolysis is assumed to have the higher resilience; hence the highest potential to improve income levels[b,d,e,j]
12	Based on the existing infrastructure for fossil technologies and forecasted requirements for new hydrogen processing, storage, and distribution, a large amount of permanent and temporary jobs will be created. This amount will depend on labor intensity of technology, as well as ease of processing/harvesting materials and energy factors. Since renewables have already a head start it is the option with the highest potential, followed by biological and steam reforming of ethanol. Coal gasification and steam reforming of natural gas are options with lowest potential [b,d,f,j]

[a]Granovskii et al. (2006)
[b]Hotza and Diniz da Costa (2008)
[c]Gaudernack and Lynum (1998)
[d]McLellan et al. (2005)
[e]Carmo de Lima and Veziroğlu (2001)
[f]IEA (2005)
[g]Ribeiro and Cardoso (2003)
[h]Afgan and Carvalho (2002)
[i]Bell et al. (2006)
[j]Galli (2011)

6.5.7 MCA Final Score and Ranking of Options

With the results of combining Tables 6.2 and 6.3, it is possible to obtain a final score for each of the proposed options. Table 6.5 summarizes the outcome of Appendix III.

The outcome of the multi-criteria analysis leans towards the implementation of hydrogen technologies based on biological sources or renewable-powered electrolysis. However, it is important to analyze the variables involved throughout a sensitivity analysis to provide a measure of robustness and indicate whether the used variables cover uncertainties adequately.

Table 6.4 Assigned values of indicators based on "ideal realistic values"

1. Potential to reduce GHG and Toxic emissions (%)	0–19 %	20–39 %	40–59 %	60–79 %	80–100 %
	10	30	60	80	100
2. Potential to substitute fuel source for electricity (1 = very low, 5 = very high)	Very Low (1)	Low (2)	Neutral (3)	High (4)	Very high (5)
	10	30	50	75	100
3. Land use impact (1 = very low, 5 = very high)	Very Low (1)	Low (2)	Neutral (3)	High (4)	Very High (5)
	80	65	40	25	5
4. Material and energy intensity (1 = very low, 5 = very high)	Very Low (1)	Low (2)	Neutral (3)	High (4)	Very High (5)
	100	70	50	30	5
5. Energy security and economic resilience (1 = very low, 5 = very high)	Very Low (1)	Low (2)	Neutral (3)	High (4)	Very High (5)
	20	40	60	80	100
6. Adaptability of storage and distribution (1 = very low, 5 = very high)	Very Low (1)	Low (2)	Neutral (3)	High (4)	Very High (5)
	30	50	70	80	90
7. Adaptability for industrial and vehicular use (1 = very low, 5 = very high)	Very Low (1)	Low (2)	Neutral (3)	High (4)	Very High (5)
	10	30	50	70	90
8. Required investment (1 = very low, 5 = very high)	Very Low (1)	Low (2)	Neutral (3)	High (4)	Very High (5)
	10	20	40	60	80
9. Potential to improve health (%)	0–19 %	20–39 %	40–59 %	60–79 %	80–100 %
	20	40	60	80	100
10. Potential to improve quality of life (1 = very low, 5 = very high)	Very Low (1)	Low (2)	Neutral (3)	High (4)	Very High (5)
	20	40	60	80	100
11. Potential to improve income levels (1 = very low, 5 = very high)	Very Low (1)	Low (2)	Neutral (3)	High (4)	Very High (5)
	20	40	60	80	100
12. Increase of trade specific jobs	Very Low (1)	Low (2)	Neutral (3)	High (4)	Very High (5)
	10	20	40	60	80

Table 6.5 MCA final scores and ranking

Option	MCA score	Ranking
1. Hydrogen from coal gasification with carbon capture	495	Fourth
2. Hydrogen from electrolysis powered by renewable sources	805	First
3. Hydrogen from biological process [Biophotolysis]	805	First
4. Hydrogen from steam reforming of natural gas	570	Third
5. Hydrogen from steam reforming of ethanol	705	Second

6.5.8 *Sensitivity Analysis*

The previous results show an inclination towards hydrogen produced via renewable technologies. This was most definitely an expected outcome from the analysis; however, since all indicators received a different weighting scale, it does come as a surprise that Options 2 and 3 yielded the same final score. The level of uncertainty managed with the indicators can be considered as high, as no empirical information was available to support the weighting of indicators. Assumptions, such as the imminent development of hydrogen production from electrolysis powered by renewable sources as well as from biological sources, lead to a time frame of 10–20 years for these technologies to be in an acceptable operational stage. However, it is uncertain whether such technologies will achieve the estimated functionality on that time frame.

Other uncertainties lay on the future extraction and processing amounts of fossil fuels along with international agreements and policy to restrict their use and curb GHG emissions. It is difficult to forecast to what extent international market prices will drive the implementation of hydrogen production by any of the means given previously. It is clear that hydrogen use on ICE or fuel cells will start penetrating the Brazilian market, but it is not clear as to how fast and with what intensity it is going to occur. Economic stability and energy security will mark the way to a successful implementation, but it will, most importantly, allow a smooth transition into a single technology or a combination of any.

6.6 Discussion

The MCA results clearly establish the value and ranking of green technologies against fossil-based technologies for hydrogen production. At this point it becomes important to acquire an open perspective to interpret the results and ask the right questions to obtain more than solid numbers from the analysis. The information provided by the results could be interpreted in two different ways, the first one being a strong and cutting push to start shifting all technologies towards renewable power and biological hydrogen production, which is in fact completely unrealistic. Although there is a great need to introduce hydrogen production to achieve a high score on all selected indicators, a second approach for interpreting the results is feasible and has a better grasp on reality.

This second approach lays its foundation on the fact that, in order to accomplish energy security amongst other identified benefits and drivers from hydrogen, a transition or several transition periods would need to take effect before such benefits could be redeemed. These transition periods will be marked by current technologies and those under development that require a time lapse for maturity before they can become operational and eventually be deployed within the consumer market.

Although it is true that fossil fuels are still required to satisfy part of Brazil's energy needs, much has been done to displace fossil fuels from the energy matrix for electricity generation as well as for vehicular purposes. In Brazil, a transition from fossil fuel hydrogen generation to renewable hydrogen production could be easier than for other countries. The high percentage of hydropower used for electricity generation along with standards set for gasoline-ethanol blends are the best examples and a good foundation for transitions to be laid into policy.

Hence, a set of transition periods that will enable Brazil to accomplish the aforementioned goals for energy security could be proposed as depicted in Fig. 6.7. A first transition period could be set by using Options 1, 4, and 5, thus enabling the steam reforming of ethanol to gain strength over the years through giving opportunity for maturing PEMFC, phasing out hydrogen production from coal gasification and steam reforming of natural gas. This transition period will have to push the deployment of fuel cells within the light vehicle market, thus increasing hydrogen demand and consolidating offer from the steam reforming of ethanol. While this transition period is in effect, technologies for Options 2 and 3 could have enough time for research and development maturity to the point where hydrogen could then be produced in an economic fashion from these methods. With parts of hydrogen storage and distribution infrastructure built during the first transition, the following transitions would have a smoother path towards implementation (Fig. 6.8).

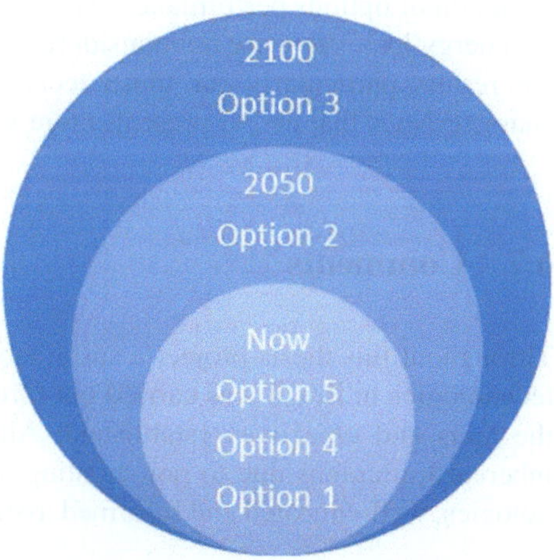

Fig. 6.8 Progression of assessed options throughout 100-year time span

A final approach could be found in using overarching indicators individually or selecting indicators depending on the project or problem to be solved. There might be instances were social indicators will function better than economic or environmental ones to tackle health or risk issues. The same will happen when trying to solve environmental issues with social indicators. The advantage of using MCA for this particular project was that the analysis is able to interchange weighting values on the weighting matrix and assign higher values to those indicators that were thought of being more important or to attain a heavier weight on the decision-making process.

The strength of indicators will be based on the type of problem being implemented. However, the validation of indicators can only be assessed when actual data covering all the options is able to be collected and weighted. Policy- and decision-makers can use overarching indicators and the MCA methodology to address medium and large size projects that could impact all pillars of sustainability, while at the same time providing acceptable results that will give each pillar and involved stakeholders an accurate qualitative value.

While revising the current study with GOSE (Group of Optimization of Energy Systems), it was inquired whether a biomass gasification option should be considered in the analysis. Given the natural occurring amounts of bagasse as a result of sugarcane extraction and ethanol production, such types of feedstock pose an attractive perspective for comparison purposes. However, in order to keep the analysis balanced by two "Full green" options (Options 2 and 3), two contrasting "Full fossil" options (Options 1 and 4) were selected along with one last option that will represent both extremes (Option 5). Adding a biomass gasification option will have broken the balance of the analysis, unless a substitution for Option 5 was selected.

It is important to note that material and energy balances for technologies under analysis were not performed due to the evident lack of empirical information. Individual material and energy balances will in fact become the main pillar for this type of analysis, indicating energy/material input–output ratio to determine resource feasibility of each option. The obtained information will directly substitute the estimated values within the weighting matrix and provide a more accurate depiction of options performance. Material and energy balances could also account for energy losses that are not considered within this analysis, and could easily shift the results particularly for those technologies that share renewable and fossil characteristics that are, at present, more viable options for implementation.

6.7 Comments

Throughout this thesis project a sustainability assessment of hydrogen production technologies in Brazil was carried out through the multi-criteria analysis, fulfilling the aims and objectives established. Although the project was developed with inherent limitations due to non-existing data for some hydrogen production technologies, well-educated and informed assumptions were taken into account in order to provide a sound analysis. Establishing a decision-making framework as well as a point of convergence for the analysis facilitated the identification and selection of both options and indicators for proper weighting, scale setting, final score, and ranking.

Final results of MCA ranked option #2 (HEPRS) and #3 (HBP) with a score of "805," placing both options in first place. Option #5 (HSRE) obtained a final score of "705," option #4 (HSRNG) a score of "570," and option #1 (HCGCC) a score of "495"; placing them on second, third, and fourth place, respectively. MCA final scores lean towards the implementation of renewable-based hydrogen production, taking into account the heavy burden of limiting storage and distribution infrastructure required to establish hydrogen within the consumer market. Although results are presented in a numeric fashion, various approaches were evaluated to interpret such and provide an acceptable outcome within a specific targeted timeframe.

Two realistic approaches were deducted from MCA results, the first one establishing a set of transition periods aimed at using policy and social tools to strengthen hybrid renewable hydrogen production through existing viable technologies, while at the same instance providing newer and promising technologies enough time to mature into the energy matrix and consumer markets. By allowing existing hydrogen production technologies to thrive, storage and distribution infrastructure will also have an opportunity to consolidate and spread throughout the consumer market. Also, an emerging demand will aid in displacing former fossil and hybrid technologies, expanding the horizon for pure renewable production sources.

A final approach is recognized by specifically selecting overarching indicators that can define a particular problem or decision-making dilemma, engulfing detailed characteristics that provide indicators with the appropriate strength and versatility for measuring one or more criteria in a qualitative or quantitative manner. It can then be concluded that the correct approach for hydrogen production in Brazil, according to this study, will depend on current technologies and the realistic development-implementation of new ones, renewable energy sources within the energy matrix, energy policy forecasting and consistency throughout 2100, focusing on phasing out fossil technologies and finally the ability of the Brazilian government to enforce established policies.

Directed sustainability should lay its foundation on breaking paradigms, but such paradigms should be approached according to the particular needs of the country in question. Economic, social, or environmental problems can all be addressed through policy-making, but ensuring the way to follow up and enforce that policy is what directed sustainability will accomplish by fitting and designing mechanisms to increase efforts towards those measures.

References

Afgan NH, Carvalho MG (2002) Multi-criteria assessment of new and renewable energy power plants. Energy 27:739–755

Azapagic A, Perdan S (2000) Indicators of sustainable development for industry: a general framework. Trans IChemE 78B:244–246

Balat M (2009) Political, economic and environmental impacts of biomass-based hydrogen. Int J Hydrogen Energy 34:3589–3603

Bell ML, Davis DL, Gouveia N, Borja-Aburto VH, Cifuentes LA (2006) The avoidable health effects of air pollution in three Latin American cities: Santiago, São Paulo, and Mexico City. Environ Res 100(3):431–440

Brazil Works (2012) Electricity generation in Brazil: myths and preconceptions about the construction of Belo Monte. http://www.brazil-works.com/wp-content/uploads/2012/10/Myths-and-Preconceptions-about-Belo-Monte1.pdf. Accessed 8 May 2013

Carbon Counts Company Ltd (2010) CCS roadmap for industry: High-purity CO_2 sources. Sector assessment—final draft report. http://cdn.globalccsinstitute.com/sites/default/files/publications/15686/ccs-roadmap-industry-high-purity-co2-sources-sectoral-assessment.pdf. Accessed 5 May 2013

Carmo de Lima L, Veziroğlu TN (2001) Long-term environmental and socio-economic impact of a hydrogen energy program in Brazil. Int J Hydrogen Energy 26:39–45

Carrera G, Mack A (2010) Sustainability assessment of energy technologies via social indicators: Results of a survey among European energy experts. Energy Policy 38:1030–1039

Cerqueira Leite RC, Leal MRLV, Cortez LAB, Griffin WM, Scandiffio MIG (2009) Can Brazil replace 5% of the 2025 gasoline world demand with Ethanol? Energy 34:655–661

Dincer I (2007) Environmental and sustainability aspects of hydrogen and fuel cell systems. Int J Energy Res 31:29–55

Ministerio de Minas e Energia (2012) Balanço Energético Nacional (National Energy Balance), Empresa de Pesquisa Energética. https://ben.epe.gov.br/downloads/S%C3%ADntese%20do%20Relat%C3%B3rio%20Final_2012_Web.pdf. Accessed 17 Apr 2013

Food and Agricultural Organization (FAO) (2011) The global bioenergy partnership sustainability indicators for bioenergy. http://www.globalbioenergy.org/fileadmin/user_upload/gbep/docs/Indicators/The_GBEP_Sustainability_Indicators_for_Bioenergy_FINAL.pdf. Accessed 25 April 2013

USDA Foreign Agricultural Service (2012) Brazil biofuels annual report 2012. http://gain.fas.usda.gov/Recent%20GAIN%20Publications/Biofuels%20Annual_Sao%20Paulo%20ATO_Brazil_8-21-2012.pdf. Accessed 24 March 2013

Fuel Cell & Hydrogen Energy Association (FCHEA) (n.d.) Hydrogen Production from Coal. http://www.fchea.org/core/import/PDFs/factsheets/Hydrogen%20Production%20From%20Coal_NEW.pdf. Accessed 30 Apr 2013

Galli E (2011) Frame analysis in environmental conflicts: the case of ethanol production in Brazil. Doctoral thesis in Industrial Ecology, KTH Industrial Engineering and Management, Division of Industrial Ecology. US AB, Sweden

Gamper CD, Thöni M, Weck-Hannemann H (2006) A conceptual approach to the use of cost benefit and multi criteria analysis in natural hazard management. In: Natural Hazards and Earth System Sciences. http://www.nat-hazards-earth-syst-sci.net/6/293/2006/nhess-6-293-2006.pdf. Accessed 8 April 2013

Gaudernack B, Lynum S S (1998) Hydrogen from natural gas without release of CO_2 to the atmosphere. Int J Hydrogen Energy 23(12):1087–1093

Granovskii M, Dincer I, Rosen MA (2006) Life cycle assessment of hydrogen fuel cell and gasoline vehicles. Int J Hydrogen Energy 31:337–352

Hirschfeld J, Hadley D, Mongruel R, D'Hernoncourt J (2011) Multi-criteria analysis—specification sheet and supporting material. In: Spicosa project report, IOeW, Berlin

Hotza D, Diniz da Costa JC (2008) Fuel cells development and hydrogen production from renewable resources in Brazil. Int J Hydrogen Energy 33:4915–4935

Hydroxsystems (2013) Technology. http://hydroxsystems.com/page/11/hho-technology.html. Accessed 25 April 2013

International Rivers (2012) The Brazilian electricity sector and sustainability in the 21st century: opportunities and challenges. http://www.internationalrivers.org/node/7525. Accessed 6 May 2012

International Energy Agency (2005) Prospects for hydrogen and fuel cells, energy technology analysis. International Energy Agency—Head of Publication Service, Paris, France

Lèbre LRE, Santos PA, Simões AF (2011) Biofuels and sustainable energy development in Brazil. World Dev 39(6):1026–1036

Mann M, Chornet E, Czernik S, Wang D (n.d.) Biomass to hydrogen via pyrolysis and reforming, National Renewable Energy Laboratory. http://web.anl.gov/PCS/acsfuel/preprint%20archive/Files/39_4_WASHINGTON%20DC_08-94_1034.pdf. Accessed 4 Apr 2013

Mattos MC (1985) National hydrogen energy program in Brazil. Int J Hydrogen Energy 10 (9):601–606

McLellan B, Shoko E, Dicks AL, Diniz da Costa JC (2005) Hydrogen production and utilization opportunities for Australia. Int J Hydrogen Energy 30:669–679

Molburg JC, Doctor RD (2003) Hydrogen from steam-methane reforming with CO_2 capture. Argonne National Laboratory. In: 20th Annual International Pittsburgh Coal Conference. http://www.netl.doe.gov/

Momirlan M, Veziroglu TN (2002) Current status of hydrogen energy. Renew Sustain Energy Rev 6:141–179

BBC News (2013, February 16) Brazil's Marina Silva launches "sustainability party"

Palma V, Castaldo F, Ciambelli P, Iaquianiello G (2012) Hydrogen production through catalytic low-temperature bio-ethanol steam reforming. Clean Technol Environ Policy 14:973–987

Parris RM, Kates RW (2003) Characterizing and Measuring Sustainable Development. Annu Rev Environ Resour 28:559–586

Ribeiro H, Cardoso MRA (2003) Air pollution and children's health in São Paulo (1986–1998). Soc Sci Med 57:2013–2022

Silveira JL, Braga LB, de Souza ACC, Antunes JS, Zanzi R (2009) The benefits of ethanoluse for hydrogen production in urban transportation. Renew Sustain Energy Rev 13:2525–2534

Soccol CR, Vandenberghe LPS, Costa B, Woiciechowski AL, de Carvalho JC, Madeiros ABP, Francisco AM, Bonomi LJ LJ (2005) Brazilian biofuel program: an overview. J Sci Ind Res 64:897–904

Sordi A, Silva EP, Milanez LF, Lobkov DD, Souza SNM (2009) Hydrogen from biomass gas steam reforming for low temperature fuel cell: energy and exergy analysis. Braz J Chem Eng 26(1):159–169

Srirangan K, Pyne ME, Chou CP (2011) Biochemical and genetic engineering strategies to enhance hydrogen production in photosynthetic algae and cyanobacteria. Bioresour Technol 102:8589–8604

The Bioenergy Site (2013) Brazil faces looming transport fuel shortfall. http://www.thebioenergysite.com/news/12573/brazil-faces-looming-transport-fuel-shortfall. Accessed 18 April 2013

The Guardian (2013a) Carbon bubble will plunge the world into another financial crisis, 19/04/2013

The Guardian (2013b) Brazil's green flag-bearer Marina Silva ready to get back in the race, 22/04/2013

The Guardian (2013c) Belo Monte dam construction halted by Brazilian court, 08/16/2012

Turcksin L, Macharis C, Lebeau K, Boureima F, Van Mierlo J, Bram S, De Ruyck J, Mertens L, Jossart JM, Gorissen L, Pelkmans L (2011) A multi-actor-multi-criteria framework to assess the stakeholder support for different biofuel options: the case of Belgium. Energy Policy 39:200–214

Turner JA (2004) Sustainable hydrogen production. Science 305:972–974

World Commission on Environment and Development (1987) Our common future. http://www.un-documents.net/wced-ocf.htm. Accessed 16 April 2013

Chapter 7
Conclusions

José Luz Silveira

Based on the results and data available in literature, it can be concluded that there is no optimal hydrogen production process from energy, ecological and economic points of view at the same time. What should better define the most appropriate process is the scenario that exists around the installation.

With the thermodynamic efficiency analysis, the values obtained for the steam reforming processes present a slight variation between them. Ethanol reforming is the most efficient process from an energy point of view, due to operating at lower temperatures (400 °C). The difference in efficiency between the steam reforming of natural gas and biogas is influenced by the amount of methane in the composition adopted for natural gas (90 %) and for biogas (75.7 %), thus the steam reforming of natural gas is more efficient than the steam reforming of biogas.

Hydrogen production costs decrease as the capital amortization period increases, with decreasing the annual interest rate and increasing equivalent period of use. Most processes involve a high initial investment. The photobiological process is an exception because it does not require a high initial investment. However, there must be constant acquisition of algae growth medium (TAP) and hydrogen production (TAP-S), which makes its operation costly.

From an ecological efficiency perspective, the steam reforming of biogas, considering the amount of methane that would be released, and the photobiological process using the alga *C. reinhardtii* stand out. For such processes, it was considered 100 % of ecological efficiency, whereas the value of (CO_2)e of both would be negative, given the associated carbon credit.

J.L. Silveira (✉)
Group of Optimization of Energy Systems—GOSE, College of Engineering of Guaratinguetá, Institute of Bioenergy Research—IPBEN, São Paulo State University—UNESP, Dr. Ariberto Pereira da Cunha Ave, 333. Pedregulho, Guaratinguetá 12516-410, SP, Brazil
e-mail: joseluz@feg.unesp.br

J.L. Silveira (ed.), *Sustainable Hydrogen Production Processes*, Green Energy and Technology, DOI 10.1007/978-3-319-41616-8_7

Generally speaking, all hydrogen production processes showed high ecological efficiency, since not only emissions are considered, but also the energy efficiency; moreover, they can even be influenced by the fuel's lower heating value (in the case of steam reforming processes).

Among the cases studied, without incorporating the carbon credit costs, the steam reforming process of ethanol stand out, which has a high energy efficiency and lower hydrogen production costs.

With the incorporation of carbon credits, the steam reforming of biogas, whose technology, for being similar to that used in natural gas reforming, is well known and developed. Hydrogen production from algae is also feasible, but it still requires further studies on large-scale production establishment.

The following topics are being studied by researchers on the following subjects:

- Study of the sustainability of hydrogen production processes
- Thermoeconomic study (exergy + engineering economics + thermodynamics optimization) on hydrogen production from steam reforming processes
- Including the cost of the purificator used in steam reforming processes in engineering economics calculations, and reduce the cost of electrolysis processes by selling the produced oxygen
- Including the cost associated with storage in the calculations of hydrogen production costs, which require investment and expenditures on electricity
- Considering the amount of energy spent on hydrogen storage in the thermodynamic efficiency calculations
- Including the emissions from the purificator and hydrogen storage in the ecological efficiency study
- Studying the possibilities of using hybrid systems of hydrogen production
- Including the carbon credit gained in the processes of steam reforming of biogas and photobiological hydrogen production from algae in any calculations of engineering economics

Appendix A: Tables Showing Gibbs Free Energy as a Function of Temperature of Formation Reactions

Table A.1 Gibbs free energy of methane formation at different temperatures (Adapted from David 2012)

$C + 2H_2 \rightarrow CH_4$	
T (K)	ΔG (kJ/mol)
298.15	−50.53
300	−50.381
400	−41.827
500	−32.525
600	−22.69
700	−12.476
800	−1.993
900	8.677
1000	19,475
1100	30.358
1200	41.294
1300	52.258
1400	63.231
1500	74.2

Table A.2 Gibbs free energy of ethane formation at different temperatures (Adapted from David 2012)

$2C + 3H_2 \rightarrow C_2H_6$	
T (K)	ΔG (kJ/mol)
298.15	−32.015
300	−31.692

(continued)

J.L. Silveira (ed.), *Sustainable Hydrogen Production Processes*, Green Energy and Technology, DOI 10.1007/978-3-319-41616-8

Table A.2 (continued)

$2C + 3H_2 \rightarrow C_2H_6$	
T (K)	ΔG (kJ/mol)
400	−13.473
500	5.912
600	26.086
700	46.8
800	67.887
900	89.231
1000	110.75
1100	132.385
1200	154.096
1300	175.85
1400	197.625
1500	219.404

Table A.3 Gibbs free energy of ethanol formation at different temperatures (Adapted from David 2012)

$2C + 3H_2 + 1/2O_2 \rightarrow C_2H_5OH$	
T (K)	ΔG (kJ/mol)
298.15	−167.874
300	−167.458
400	−144.216
500	−119.82
600	−94.672
700	−69.023
800	−43.038
900	−16.825
1000	9.539
1100	36
1200	62.52
1300	89.07
1400	115.63
1500	142.185

Table A.4 Gibbs free energy of carbon dioxide formation at different temperatures (Adapted from David 2012)

$C + O_2 \rightarrow CO_2$	
T (K)	ΔG (kJ/mol)
298.15	−394.373
300	−394.379
400	−394.656
500	−394.914
600	−395.152

(continued)

Table A.4 (continued)

$C + O_2 \rightarrow CO_2$	
T (K)	Δ*G* (kJ/mol)
700	−395.367
800	−395.558
900	−395.724
1000	−395.865
1100	−395.984
1200	−396.081
1300	−396.159
1400	−396.219
1500	−396.264

Table A.5 Gibbs free energy of carbon monoxide formation at different temperatures (Adapted from David 2012)

$C + 1/2O_2 \rightarrow CO$	
T (K)	Δ*G* (kJ/mol)
298.15	−137.168
300	−137.333
400	−146.341
500	−155.412
600	−164.48
700	−173.513
800	−182.494
900	−191.417
1000	−200.281
1100	−209.084
1200	−217.829
1300	−226.1518
1400	−235.155
1500	−243.7424

Table A.6 Gibbs free energy of water formation from carbon at different temperatures (Adapted from David 2012)

$H_2 + 1/2O_2 \rightarrow H_2O$	
T (K)	Δ*G* (kJ/mol)
298.15	−228.582
300	−228.5
400	−223.9
500	−219.05
600	−214.008
700	−208.814
800	−203.501

(continued)

Table A.6 (continued)

$H_2 + 1/2O_2 \rightarrow H_2O$	
T (K)	ΔG (kJ/mol)
900	−198.091
1000	−192.603
1100	−187.052
1200	−181.45
1300	−175.807

Appendix B: Adaptation of Cardu and Baica's Methodology

As it is proposed by Cardu and Baica (1999), it is assumed that the ecological efficiency has the following form:

$$\varepsilon = [c \cdot \varphi(\eta, \textit{indicador}) \cdot \psi(\textit{indicador})]^n$$

Constant "c" and the exponent "n" are going to be calculated by the boundary conditions. The Pollutant Indicator is used to quantify the environmental impact of a technology. Its unit is $kgCO_{2(eq)}/kgH_2$.

By following the concepts of the authors, it is proposed that $\varphi(\eta, \textit{indicator})$ is as follows:

$$\varphi = \frac{\eta_{\text{system}}}{(\eta_{\text{system}} + \textit{indicator})}$$

where η_{system} is the efficiency of the hydrogen production system.

In order to reduce the range of values and approximate the efficiency curves of various technologies, the authors propose using function $\psi(\textit{indicator})$ as $\psi = \ln(K - \textit{indicator})$. As the logarithm transforms a high number into a lower number, the distance between extreme points is attenuated. For example, ln 10 = 2.3 and ln 100 = 4.6. Therefore, an interval of [10–100] became [2.3–4.6].

In order to validate the equation, three boundary conditions are adopted, which are different from those defined by Cardu and Baica and with the indicator in $kgCO_{2(eq)}/kgH_2$:

Condition 1. If the indicator = 0, $\varepsilon = 1$ for any η.

Condition 2. If the indicator = 50 $kgCO_{2(eq)}/kgH_2$, $\varepsilon = 0$ for any η. This indicator corresponds to the impact of Lignite (coal with high carbon content).

J.L. Silveira (ed.), *Sustainable Hydrogen Production Processes*, Green Energy and Technology, DOI 10.1007/978-3-319-41616-8

Condition 3. In this case, the hydroelectric power plant was adopted because the ecological efficiency of this hydrogen production process has already been previously calculated by Braga (2010), Siqueira and Silveira (2011), which corresponds to 0.99. It is obtained the indicator = 0.34 $kgCO_{2(eq)}/kgH_2$ and $\eta = 0.78$ (value calculated for the hydrogen production process with electricity from a hydroelectric power plant).

If:

$$\psi = \ln(K - indicator)$$

Thus, through condition 2 and formulation 1, it is obtained:

$$\psi = \ln(51 - indicator)$$

consequently:

$$\varepsilon = \left[c \cdot \frac{\eta_{\text{system}}}{(\eta_{\text{system}} + indicator)} \cdot \ln(51 - indicator)\right]^n$$

From condition 1, it is found that
$1 = [c \cdot \ln(51)]^n$
Then $c = 0.25$
From condition 3, it is found that $n = 0.023$ for $\varepsilon = 0.99n$ and the ecological efficiency equation for hydrogen production process is written as:

$$\varepsilon = \left[0.25 \cdot \frac{\eta_{\text{system}}}{(\eta_{\text{system}} + indicator)} \cdot \ln(51 - indicator)\right]^{0.023}$$

where:
Indicator: $[kgCO_{2(eq)}/kgH_2]$

References

Braga LB (2010) Análise econômica do uso de célula a combustível para acionamento de ônibus urbano. Dissertation—Curso de Engenharia Mecânica, Departamento de Energía, UNESP, Guaratinguetá

Cardu M, Baica M (1999) Regarding a global methodology to estimate the energy-ecologic efficiency of thermopower plants. Energy Convers Manage 40:71–87

David R (2012) CRC handbook of chemistry and physics, 87th edn. Internet version 2007

Siqueira RBP, Silveira JL (2011) Eficiência ecológica aplicada a uma PCH em função da operação de um reservatório hipotético. PCH Notícias e SHP News 50:24–28

Index

J.L. Silveira (ed.), *Sustainable Hydrogen Production Processes*, Green Energy and Technology, DOI 10.1007/978-3-319-41616-8

MIX
Papier aus verantwortungsvollen Quellen
Paper from responsible sources
FSC® C105338

If you have any concerns about our products,
you can contact us on
ProductSafety@springernature.com

In case Publisher is established outside the EU,
the EU authorized representative is:
Springer Nature Customer Service Center GmbH
Europaplatz 3, 69115 Heidelberg, Germany

Printed by Libri Plureos GmbH
in Hamburg, Germany